AFRICA'S INLAND FIS

Fountain Publishers Ltd
P.O. Box 488
Kampala - Uganda
E-mail:fountain@starcom.co.ug
Web:www.fountainpublishers.com

First published 2002

The findings of this publication do not necessarily reflect the opinions or policies of the EU, the EDF, the LVFRP, FIRRI, KMFRI, TAFIRI or any other institution with which it is associated, or which is mentioned in the text.

ISBN 9970 02 293 8

Cataloguing-in-Publication Data

Geheb, Kim (ed).
Africa's inland fisheries: the management challenge/Kim Geheb (and) Marie Theresa Sarah – Kampala: Fountain, 2002
p. cm
Includes Index.
ISBN 9970-02-293-8

1. Management and Development of Fisheries – Africa
2. Fisheries Resources – Africa I. Title
   338.3727'096 – dc21

Africa's Inland Fisheries

The Management Challenge

Edited by

Kim Geheb & Marie-Therese Sarch

Fountain Publishers

Fountain Publishers Ltd
P.O. Box 488
Kampala - Uganda
E-mail:fountain@starcom.co.ug
Web:www.fountainpublishers.com

First published 2002

The findings of this publication do not necessarily reflect the opinions or policies of the EU, the EDF, the LVFRP, FIRRI, KMFRI, TAFIRI or any other institution with which it is associated, or which is mentioned in the text.

ISBN 9970 02 293 8

Cataloguing-in-Publication Data

Geheb, Kim (ed).
Africa's inland fisheries: the management challenge/Kim Geheb (and) Marie Theresa Sarah – Kampala: Fountain, 2002
p. cm
Includes Index.
ISBN 9970-02-293-8

1. Management and Development of Fisheries – Africa
2. Fisheries Resources – Africa I. Title
338.3727'096 – dc21

Dedications

KG

to
Kim
and to my generations
Andy, Per, Maia and Kiira

MTS

To all those who work day in day out to make fisheries management succeed in Africa and above all to those whose lives depend on it

Contents

List of Figures

List of Tables

Notes on contributors

Edward H. Allison is at the School of Development Studies at the University of East Anglia in the UK. His primary research interests are in fisheries management and livelihoods, conservation and development. *Contact details: School of Development Studies, University of East Anglia, Norwich NR4 7TJ, U.K. Email: e.allison@ uea.ac.uk*

Kevin Crean is a senior lecturer at the University of Hull International Fisheries Institute. His research interests include the development and management of artisanal fisheries, boundary institutions, the fisheries of the European Union and strategies for devolved management of fisheries. *Contact details: University of Hull International Fisheries Institute, Cottingham Road, Hull HU6 7RX, U.K. Email: k.crean@hull.ac.uk*

Steve J. Donda is a Fisheries Socio-economist with the Malawi Department of Fisheries and is based in Lilongwe. His primary research areas and interests lie in participatory fisheries management or co-management, and how local institutions can be utilised to initiate the introduction of co-management. Other interests include how research results can be used to formulate project interventions in co-management activities. *Contact details: Department of Fisheries, P.O. Box 593, Lilongwe, Malawi. Email: sdonda@malawi.net or steve–donda@hotmail.com*

Frank Ellis is a Professor at the School of Development Studies at the University of East Anglia in the UK. His research interests lie in the fields of rural livelihoods and poverty reduction. *Contact details: School of Development Studies, University of East Anglia, Norwich NR4 7TJ, U.K. Email: f.ellis@uea.ac.uk*

Kim Geheb is a geographer trained at the University of Sussex, UK, and Indiana University, in the USA. Until December, 2001, he worked with a British-based consultancy group, UNECIA Ltd, on the EDF-funded Lake Victoria Fisheries Research Project, based in Jinja, Uganda. His main research interests are the design and implementation of natural resource management systems that include communities of resources users in their structure. *Contact details: P. O. Box 49288, Nairobi, Kenya. Email: daktari@source.co.ug*

Mafaniso M. Hara is with the Programme for Land and Agrarian Studies at the School of Government at the University of the Western Cape. His primary research interests are subsistence fisheries, (co-)management in Malawi, South Africa and Southern Africa more generally. *Contact details: Programme for*

Land and Agrarian Studies, School of Government, University of the Western Cape, P/Bag X17 Bellville 7535, Republic of South Africa. Email: mhara@uwc.ac.za

Mattias Krings holds an MA, and is based at the Institute of Historical Ethnology at the University of Frankfurt. His main research interests are drawn from Northern Nigeria, in the fields of the anthropology of migration and Hausa studies. *Contact details: Grueneburgplatz 1, 60323 Frankfurt/Main, Germany. Email: Krigs@em.uni-frankfurt.de*

Mercy Kyangwa is employed at the Fisheries Resources Research Institute in Jinja, Uganda and is, at the same time, completing a Master's thesis at Moi University in Kenya. The latter sets out to identify possible roles for fishing communities in the management of Uganda's Lake Victoria fisheries. *Contact details: FIRRI, P. O. Box 343, Jinja, Uganda. Email: kyangwa70@yahoo.com*

Ossi Veikko Lindqvist is Professor of Applied Zoology at the University of Kuopio, Finland. His primary research interests lie in the fields of fisheries and aquaculture management, development studies and science policies. He has been involved as a scientist and advisor to several fisheries and aquaculture development projects in Africa and South-East Asia. Throughout the 1990s he was both a planner and Scientific Co-ordinator of the FAO/UN Lake Tanganyika Research Project. He has been involved in numerous development issues, both nationally and internationally, and has represented Finland in the UNESCO Man and Biosphere Programme. *Contact details: Institute of Applied Biotechnology, University of Kuopio, POB 1627, FIN-70211 Kuopio, Finland. Email: ossiv.lindqvist@uku.fi*

Carolyne Lwenya is a socio-economic researcher at the Kenya Marine and Fisheries Research Institute in Kisumu, Kenya. Her research interests lie in the field of developing co-managerial solutions to Lake Victoria's fisheries management problems, and gender issues on the same lake. *Contact details: KMFRI, P. O. Box 1881, Kisumu, Kenya. Email: c.gichuki@yahoo.com*

Isaac Malasha is a research fellow at the Centre for Applied Social Sciences at the University of Zimbabwe. His primary research interests lie in the analysis of natural resources management regimes with particular reference to the Lake Kariba artisanal fishery. *Contact details: University of Zimbabwe, Centre for Applied Social Sciences, P. O. Box MP167, Mount Pleasant, Harare, Zimbabwe. Email: mambwe @mweb.co.zw*

Modesta Medard was trained at the Universities of Dar es Salaam in Tanzania, and Moi University in Kenya. She is a socio-economics researcher with the Tanzania Fisheries Research Institute in Mwanza. Her primary research interests lie in the fields of community development research studies and women's issues. Modesta Medard has been involved in socio-economic studies on Lake Victoria for over ten years. *Contact details: TAFIRI, P.O.Box 475, Mwanza, Tanzania. Email: modentara@ hotmail.com*

Heimo J. Mikkola holds a PhD in Applied Zoology and Limnology, and a docentship in tropical fisheries and aquaculture, both from the University of Kuopio, Finland. He is the FAO Resident Representative Food and Agriculture in Banjul, The Gambia. His primary research interests are fisheries and aquaculture development, and he has worked extensively on the African continent since 1977. *Contact details: FAO, Private Mail Bag 10, Banjul, The Gambia. E-mail: FAO-GMB@field.fao.org*

Hannu-Pekka Olavi Mölsä is an Acting Professor at the University of Kuopio, Finland. His primary research areas/interests lie in the fields of fisheries and aquaculture management, biotechnology in aquaculture and development studies. Hannu Mölsä has worked as a university teacher and research director in several fields in fisheries and aquaculture, as Deputy Scientific Co-ordinator of the FAO/UN Lake Tanganyika Research Project, and recently as Deputy Director of the Institute of Applied Biotechnology of Kuopio University. His experience in fisheries and aquaculture development, personnel training, and project leadership has focused primarily on Eastern Africa. He maintains also close relationships with the EU DG Dev. and other international aid organisations. *Contact details: Institute of Applied Biotechnology, University of Kuopio, POB 1627, FIN-70211 Kuopio, Finland. Email: hannu.molsa@uku.fi*

Paul Onyango is a researcher with the Tanzania Fisheries Research Institute in Sota, Tanzania. His research on Lake Victoria has taken him into the various socio-economic and nutrition fields. *Contact details: TAFIRI, P. O. Box 46 Shirati Tanzania. Email: onyango–paul@hotmail.com*

Peter M. Mvula is based at the Centre of Social Research at the University of Malawi. His research interests include poverty, livelihoods, nutrition and gender. *Contact details: University of Malawi, Centre for Social Research, P.O. Box 278, Zombà, Malawi. Email: peter–mvula@hotmail.com or pmvula@chirunga. sdnp.org.mw*

Friday J. Njaya works with the Fisheries Department in Zomba, Malawi. His research interests lie in small-scale fisheries management, with an emphasis on co-management. He was involved in the implementation of co-management on Lake Malombe from 1996-1998 under the UNDP/FAO Capture Fisheries Project and Malawi-Germany Fisheries and Aquaculture Development Project

(MAGFAD). Since then, he has been involved in the co-management of Lakes Chilwa and Chiuta. *Contact details: Fisheries Department, P. O. Box 206, Zomba, Malawi. Email: Friday@hafro.is or fnjaya@clcom.net*

Momodou Njie is a Senior Fisheries Officer with The Gambian Fisheries Department. *Contact details: Fisheries Department, 6, Marina Parade, Banjul, The Gambia. Email: FAO-GMB@field.fao.org*

Kefasi Nyikahadzoi is a lecturer at the Centre for Applied Social Sciences at the University of Zimbabwe. His primary research interests lie in the field of fisheries management. *Contact details: University of Zimbabwe, Centre for Applied Social Sciences, P. O. Box MP167, Mount Pleasant, Harare, Zimbabwe. Email: knyikahadzoi@cass.org.zw*

Joash Barack Okeyo-Owour is a Senior Lecturer in the School of Environmental Studies at Moi University in Kenya. His main areas of expertise lie in ecology and biodiversity conservation, and he has considerable interests in the conservation of wetland resources of Lake Victoria. *Contact details: P.O.Box 6423, Kisumu, Kenya. Email: viredresearch@hotmail.com*

John-Eric Reynolds is a Fisheries Planning Officer with the Food and Agriculture Organization of the United Nations. His primary research interests lie in the fields of development policy and planning, socio-economics and human ecology. Eric Reynolds has worked with the Food and Agriculture Organization in various capacities since 1987. He holds a Ph.D. in Social Anthropology from the University of Washington, and has been involved in field research, staff training, fisheries and rural development activities, and project co-ordination over the past twenty-five years, particularly in northern, eastern and southern Africa. He served as Socio-economist and Fisheries Planning Advisor for the FAO/UN Lake Tanganyika Research Project in 1991-92 and again in 1997 – 2000, and now serves as the Project Manager. *Contact details: Fishery Policy and Planning Division, Food and Agriculture Organization of the United Nations, Viale delle Terme di Caracalla, 00100 Rome, Italy. Email: eric.reynolds@fao.org*

Marie-Therese Sarch is based at the Flood Hazard Research Centre at Middlesex University in the UK. Her doctoral work (at the University of East Anglia) was carried out on the Nigerian shores of Lake Chad, and her primary research interests lie in the governance and management of natural resources in sub-Saharan Africa, agricultural and fisheries development and socio-economics. *Contact details: c/o HTS Development Ltd, Thamesfield House, Boundary Way, Hemel Hempstead, Herts HP2 7SR, UK. Email: terrisarch@cs.com*

Douglas C. Wilson is a Senior Researcher at the Institute for FisheriesResearch and Coastal Community Development in Hirtshals, Denmark. His primary research interests lie in the sociology of fisheries management in both the North and the South. The main foci of his work are 1) the effective participation of stakeholders in fisheries management, 2) thesocial processes involved in the creation of a scientific knowledge base for fisheries management, and 3) the tensions between stakeholder participation and constructing and using valid science in management. *Contact details: North Sea Centre, Willemoesvej 2, P.O.Box 104, DK-9850 Hirtshals, Denmark. Email: dw@ifm.dk*

Acknowledgements

This volume is a product of the Lake Victoria Fisheries Research Project (LVFRP), funded by the European Development Fund of the European Union (Project No. 7 RPR 372).We are hugely indebted to the financial and other support provided by the Project. In particular, we thank the Project Co-ordinator, Martin van der Knaap, for agreeing to put up with this activity, and hope that the decision to place the discussion on Lake Victoria's management into the broader context of Africa's inland fisheries, will do the LVFRP proud.

We acknowledge the tremendous support provided by the Kenya Marine and Fisheries Research Institute, the Tanzania Fisheries Research Institute and the Fisheries Resources Research Institute in Uganda. In particular, we thank and salute the socio-economic teams at these institutes.

I also thank (KG) boss at UNECIA, Peter Duff, and my colleagues on the ground, Dave MacLennan and George Passiotis, for having sailed rough and calm waters with me.

The papers contained in this volume have been anonymously reviewed by Kevin Crean at the Hull International Fisheries Institute (HIFI), and Sarah Kalloch, Fellow of the Rockefeller Foundation. The Editors offer their thanks and enormous gratitude for the huge amount of work and effort they put in.

Also at HIFI, we thank Ian Cowx for technical discussions and the gigantic efforts he has made to see this book published.

Editors
January 2002

Preface

When I first heard of the plans to compile and publish African co-management experiences, I became very enthusiastic; I really found it a good idea. Later on I started to have my doubts about this undertaking as it took a longer time than expected because communication with contributors was not always easy and naturally other project activities had to go on simultaneously. Through my work on fisheries management, I have realised that though socio-economics is a relatively young science, it should be given a chance. Personally, I am from the school of stock assessment and traditional fisheries management and, therefore, I was reluctant to embrace the co-management principle, which has become the buzz-word in fisheries management during the last five years, to resolving managerial problems.

The experience in many African situations shows that things must change. I had the opportunity to be involved in socio-economics on the Lake Victoria Fisheries Research Project and the results were an eye-opener. Many communities have their own specific problems and ways of resolving them, making every situation unique. Consequently, it is impossible to propose general co-management strategies for a lake, stretch of coastline or a river. Individual communities require intensive guidance to address their own fisheries management problems.

The essays in this book clearly indicate that co-management and community involvement are central to the challenges of fisheries management in Africa. Co-management not only involves communities but also many other stakeholders such as government institutions, artisanal and industrial processors, municipalities, etc.

I am proud that the Lake Victoria Fisheries Research Project has completed this book. I wish to commend Kim and Terri (the Project's long-term socio-economist and short-term PRA-expert respectively) for their efforts and patience in putting the book together. I should like to thank the contributors to the book for accepting to share their findings. I hope that the publication of these findings will mark a new beginning of fisheries management practice in Africa. As fisheries scientists and managers, we have an obligation to explore and implement sustainable ways of exploiting Africa's fisheries resources for an ever-growing riparian and lacustrine population.

May this book contribute to a better understanding of co-management as a viable approach to fisheries management and may it also lead to new ideas, which could be applied in other parts of the continent.

Martin van der Knaap
Coordinator Lake Victoria Fisheries Research Project
29 January 2002

Abbreviations

ADMADE	Administrative Management Design for Game Management Areas
ADMARC	Agricultural Development and Marketing Corporation
AEC	Area Executive Committee
AfDB	African Development Bank
AFDP	Artisanal Fisheries Development Project
BC	Beach Committee
BMU	Beach Management Unit
BVCs	Beach Village Committees
CAMPFIRE	Communal Areas Management Programme for Indigenous Resources
CBRM	Community-Based Resource Management
CBRDA	Chad Basin Rural Development Authority
CCM	Chama cha Mapinduzi (Revolutionary party - Tanzania)
CCRF	Code of Conduct for Responsible Fisheries
CFCs	Community Fisheries Centres
CFCMC	Community Fisheries Centre Management Committee
CFMZ	Community Fisheries Management Zone
CRH	Centre de Recherche en Hydrobiologie
CIFA	Committee for Inland Fisheries of Africa
CLU	Community Liaison Unit
COMPASS	Community Partnerships for Sustainable Resource Management Systems
CPUE	Catch per unit effort
CRD	Central River Division
DANIDA	Danish International Development Agency
DEC	District Executive Committee
DEPP	Département des Eaux, Pêches et Pisciculture
DFR	Department for Fisheries Resources
DFID	Department for International Development
DGTC	Department of Game and Tsetse Control
DNPWLM	Department of National Parks and Wild Life Management
DoF	Department of Fisheries

DRC	Democratic Republic of the Congo
EAC	East African Community
EDF	European Development Fund
EU	European Union
FAMLG	Fishermen Association Marte Local Government
FAO	Food and Agricultural Organization
FD	Fisheries Department
FDU	Fisheries Development Unit
FFMP	Framework Fisheries Management Plan
FPF	Fish Processing Factory
FRCU	Fisheries Regulations and Control Unit
GEF	Global Environmental Facility
GD	Gambian Dalasis
GTNA	Gwembe Tonga Native Authority
GTZ	Germany Technical Co-operation
ICLARM	International Centre for Living Aquatic Resource Management
IMF	International Monetary Fund
IVMC	Integrated Village Management Committee
KFA	*Kapenta* Fishermen's Association
KMFRI	Kenya Marine and Fisheries Research Institute
LCFWO	Lake Chad Fishermen's Welfare Organisation
LC	Local Council
LEU	Local Enforcement Units
LFC	Local Fisheries Council
LGA	Local Government Area
LIFE	Living in a Finite Environment
LMCs	Landing Management Committee
LTBP	Lake Tanganyika Biodiversity Project
LTFMP	Lake Tanganyika Fisheries Monitoring Programme
LTR	Lake Tanganyika Research
LVFO	Lake Victoria Fisheries Organisation
LVFS	Lake Victoria Fisheries Service
LVFRP	Lake Victoria Fisheries Research Project
MAGFAD	Malawi Germany Fisheries and Aquaculture Development

MALDECO	Malawi Development Corporation
MBC	Malawi Broadcasting Corporation
MCS	Monitoring, control, and surveillance
MNAECC	Ministry of National Affairs Employment Creation and Co-operatives
MP	Member of Parliament
MPU	Marine Patrol Unit
MRFC	Malawi Rural Finance Company
NARMAP	National Aquatic Resource Management Programme
NFC	National Fisheries Council
NGO	Non-governmental organisation
ODA	Overseas Development Administration
PA	Protected area
PFMP	Participatory Fisheries Management Programme
RoU	Republic of Uganda
SAP	Strategic Action Programme
SADC	Southern African Development Co-operation
SEDAWOG	Socio-Economic Data Working Group
SCIP	South Chad Irrigation Project
SFLP	Sustainable Fisheries Livelihoods Programme
SLA	Sustainable Livelihoods Approach
TAFIRI	Tanzania Fisheries Research Institute
TFE	Tweyambe Fishing Enterprise
TREFIP	Lake Tanganyika Regional Fisheries Programme
UNDP	United Nations Development Programme
UNICEF	United Nations International Children's Fund
UNOPS	United Nations Office for Project Services
USAID	United States Agency for International Development
WWF	World Wildlife Fund
YFCS	Yimbo Fishermen's Co-operative Society
ZMC	Zonal Management Committee
ZZSFP	Zambia-Zimbabwe SADC Fisheries Project

1
Introduction: Meeting the Challenge

Kim Geheb and Marie-Therese Sarch

Despite severe droughts and a doubling in population since the 1960s, African fish consumption has increased by one quarter (FAO, 1996b). Africa's inland fisheries have played an important role in meeting this demand. In absolute terms, inland fisheries production has soared from 250,000 t in 1950 to almost 2,000,000 t in 1999. As a proportion of total catches, inland fisheries landings have grown from less than 25% in 1951 to 49% in 1999 (FAO, 1996b, 2000).

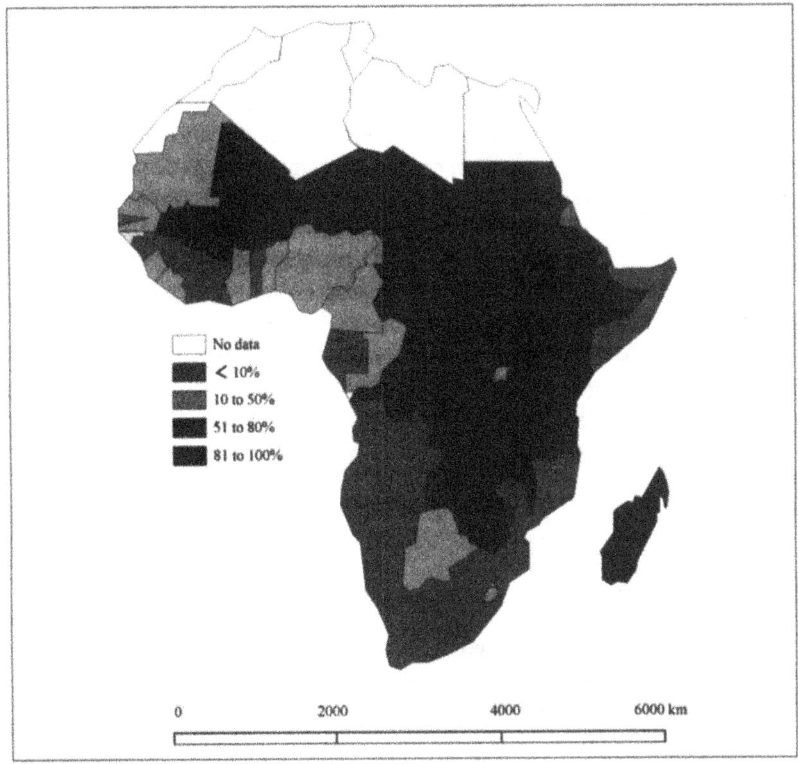

Figure 1: Inland fish production as a proportion of fish supply available per caput in sub-Saharan Afica, 1994
(Adapted from FAO, 1996b)

Inland fisheries now produce the majority of fish consumed in many African countries and almost all of that consumed in Mali, Chad and East Africa (Figure 1). For many countries, these outputs represent not only important food sources, but also vital exports - in Uganda, for example, fish exports are the country's second largest export after coffee. In addition, Africa's inland fisheries are also a vital source of employment and income for resource poor families. They are exploited almost entirely by artisanal fishing communities in predominantly rural areas. On a continent plagued by poverty, erratic political and economic systems and often depressing livelihood conditions, the role which Africa's inland fisheries have in providing alternative and often counter-seasonal sources of income is vital (Davies, 1996; Geheb, 1997; Sarch, 1999). While increasing demand for fish may well be a motivation to attract entrants into fisheries, this migration is in equal measure a reflection of poor conditions in other sectors of Africa's economy. The attraction of fisheries under these conditions lies in their open or near-open access and the relatively low level of investment required to earn at least something to eat.

The outcome of these trends is, perhaps, predictable. In 1996, the FAO estimated that artisanal fishing effort on the continent had doubled in the past decade and that most freshwater fisheries were intensively exploited. This is made nowhere more clearly than in the papers contained in this volume. The FAO goes on, '[a]s fishing effort continues to respond to the growing demand for fish, proper inland fisheries management is becoming more and more urgent.' (FAO, 1996b: 10-36)

What 'proper fisheries management' comprises is hugely debatable. It has usually meant management for equilibrium production targets such as maximum sustainable yield, with measures to achieve these targets enforced by the state (e.g. for Lake Malawi, Tweddle and Magasa, 1989; FAO, 1993; Government of Malawi, 1999). However, centralised fisheries management strategies on the continent, like equivalent systems in the North, yield little evidence of actually working, particularly in environments characterised by low levels of funding, staff, expertise and technology. It is deeply worrying that on the African continent, it is not only that the necessary *context* for the adequate functioning of centralised management systems is absent, but also that the internal machinations of these systems appear to be flawed.

As a result, new management styles are being developed to achieve a range of management objectives. Many of these advocate increased roles for communities of resource users in their structure. A recent GTZ initiative examines how the management of traditional fisheries can be enhanced to increase their production (COFAD, 2002). Amongst the attractions of such management styles are that they reduce the costs of management, improve monitoring of the resource, are democratic, and promise greater regulatory enforcement than do centralised, state-based management strategies. Perhaps more importantly, in a domain where it is often difficult to perceive the impact of the state's role in the management loop, these alternative management systems represent potential strategies for managerial reform

As a result, the discussion on fisheries management in Africa shows an increasing interest in community and co-management strategies (e.g. Normann, Nielsen and Sverdrup-Jensen, 1998). These approaches are, however, often based on unjustified assumptions about equilibrial fish stocks and livelihoods based entirely on fishing. These assumptions lead to the uncritical promotion of territorial use rights in undifferentiated and idealised constructs of a 'community' united by fishing interests (Sarch and Allison, 2000; Agarwal and Gibson, 1999; Malasha, this volume). The assumption in both cases is that fish yields can be both optimised and stabilised by better management.

Using fish stocks as the unit around which management is based is highly problematic. The determinants of fisheries exploitation find their roots in the livelihoods systems communities of resource users develop. Typically, fisheries comprise only a part of the portfolio of strategies people employ to ensure that livelihood goals are met (cf. Davies, 1996; Geheb, 1997; Sarch, 1999; Sarch and Allison, 2000). Such livelihood systems seek to ensure that rural peoples always have a variety of options on which they can draw, and mechanisms that they can utilise to minimise risks and ameliorate vulnerability. It is the dynamics of such systems that determine fisheries exploitation, and fisheries management cannot afford to ignore the ways in which these work.

Interventions designed to achieve better management of Africa's inland fisheries have been initiated in many of Africa's inland fisheries over the past decade. This volume brings together experiences of diverse management initiatives in the inland fisheries of East, Southern and Western Africa. The models used to achieve better management vary, as do the fisheries to which they are applied, and there is little consensus on an appropriate model for managing Africa's inland fisheries. In bringing together the experiences of a range of African fisheries management initiatives, this volume aims to illuminate the dilemmas that continue to face Africa's fisheries managers:

(a) What are the reasons for the failures of state-based management systems to achieve optimal yields from their inland fisheries? Can they be attributed to a scarcity in the human and financial resources allocated to them, an absence of the political will to fully implement such systems, or are there fundamental flaws in the central state-based management model?

(b) Where Africa's customary and community-based management systems have been able to conserve fish stocks and sustain fishing livelihoods, can they continue to do so in the face of the increasing intensification of fisheries exploitation?

(c) What does 'co-management' mean for Africa's inland fisheries? What can be learned from attempts to foster co-operation between fishing communities and state fisheries managers? What changes does co-management require from fisheries managers and from fishing communities? Can fisheries managers and fishing communities share an understanding of a fishery and the rules that define its management system? Do they? Is co-management a viable option?

(d) Should fisheries managers manage fish or humans? Can fisheries management be understood through the study of fisheries alone? To what extent is the success or otherwise of fisheries management determined by factors outside the fishery?

This volume sets out to attempt answers to the above questions. Starting in Africa's south, its eleven case studies advance north and west to finish on the River Gambia. The volume concludes with a synthesis of the answers that can be derived from the case studies presented here.

The case studies

These case studies have in common, to varying degrees, the need for alternative management systems to be adopted throughout the African continent's fisheries. In many cases, the impoverished context to these fisheries is emphasised not only to demonstrate their importance in the provision of employment and nutrition, but also to explain their high levels of exploitation. Throughout this volume, it is clear that how fisheries are exploited and managed is never far from political struggles at both local and international levels.

Lake Malawi

Lake Malawi is the third largest of Africa's Great Lakes covering 30,800 km^2, of which 79% falls within Malawian territory and the remainder in Mozambian confines. Tanzania claims 5,569 km^2 of the lake's surface area. The lake has an average depth of 426 m (Vanden Bossche and Bernacsek, 1990). Drawing on their work from Malawi, Allison and his colleagues (Chapter 4) narrate how fisheries policy in Malawi is by no means static, but may be tugged in various directions by competing interests, be they a president's concerns for the profitability of his fishing businesses, over-arching national development policy, or often conflicting and changeable donor demands and policies. The paper argues that part of the reason why Malawi's national fisheries policy is, at times, weak and ambiguous may be because of (a) attempts to please and attract donors and/or (b) because the state is, in fact, very reluctant to see the wholesale devolution of powers to fishing communities. This political manoeuvring typically ensures that communities of resource users and other social concerns are often sidelined.

Lakes Chilwa, Chiuta, Malombe and the Upper Shire River

Lake Chilwa, like Lake Malawi, sits on the frontier between Malawi and Mozambique. Its surface area is very variable, but is an average of 750 km^2, of which no more than 29 km^2 lies in Mozambique. Its average depth is 2 m (Vanden Bossche and Bernacsek, 1990). Of Lake Chiuta's surface area of 200 km^2, just 40 km^2 lies across the Malawi border in Mozambique. Its depth is 5 m. Lake Malombe is even shallower, with an average depth of 4 m. This lake covers 390 km^2, all of which lies within Malawi (Vanden Bossche and Bernacsek, 1990). The Shire River commences its southwards journey from Lake Malawi, and ultimately joins the Zambezi River 520 km later. The Upper Shire is that which lies between Lake Malawi and Lake Malombe.

These four water bodies are at the centre of a remarkable management experiment, described in Friday Njaya's paper, (Chapter 2) and analysed further by Hara and his colleagues (Chapter 3). The Chilwa/Chiuta fisheries have, at the centre of their management, a community-inspired initiative aimed largely at repelling fishing gear introduced by migrant fishermen. This sets these fisheries apart from the Lake Malombe/Upper Shire River initiative that was introduced through donor-funded Fisheries Department initiatives. Both the papers discussing these management initiatives argue that this difference is the fundamental flaw in the co-management system operating on Malombe and the Upper Shire.

These two papers raise important points for debate, including an extension of Allison *et al.*'s concerns for the patchwork of donor policies and objectives that serve to influence – and, in this case, undermine – fisheries management outcomes. Hara *et al.* consider the very thorny problem of co-management representing a component of a wider process of democratisation, and wonder how communities of resource users used to generations of (relatively) authoritarian power structures will cope with these new democratic systems (also a worry raised in Geheb *et al.* in Chapter 8). Hara *et al.* also show how fisheries and the co-managerial structures developed to govern them are pulled in different directions as competing political interests seek to ensure that these institutions serve them, a process also considered in Medard *et al.*'s paper from Lake Victoria (Chapter 10).

Lake Kariba

The political overtones of fisheries resource management are not diminished on Lake Kariba, where Nyikahadzoi describes the remarkable dynamics of Zimbabwe's fisheries resource management policies that appear to have nothing remotely to do with fisheries conservation. (Chapter 5) on the northern shore of the lake, Malasha reveals the more localised political struggles caused by a new (and externally motivated) co-managerial system based on the creation of water-based territories for resource users to control. (Chapter 6) in much the same way as described by Hara *et al.*, Medard *et al.* and Njaya, it is clear that the co-managerial process in Africa, in as much as it represents the development and evolution of new managerial institutions, must be evaluated in terms of political economy.

Zambia controls 45% of Lake Kariba's total surface area of 5,364 km^2, while Zimbabwe controls the remainder. Its average depth is 29 m (Vanden Bossche and Bernacsek, 1990).

Lake Tanganyika

The management of Africa's second largest Great Lake is discussed in Reynolds *et al.*'s paper. (Chapter 7) like so many of Africa's water bodies, Tanganyika has been used as a natural feature to separate countries, augmenting considerably the management problems faced by this fishery. Its 32,900 km^2 are shared between the Democratic Republic of the Congo (DRC) (45%), Tanzania (41%), Burundi (8%) and Zambia (6%). It is also a very deep lake with a mean depth of 700 m (Vanden Bossche and Bernacsek, 1990). Reynolds *et al.*'s paper is the first of two in this volume to actually propose a management plan for the fisheries they deal with (Geheb *et al.* being the second). Throughout the paper, Reynolds and his colleagues suggest a series of administrative units for the lake and identify their responsibilities. In each case, potential benefits and possible problems are considered. In particular, the authors consider the possibility that improving the management of the fishery may improve its relative value, and hence attract new entrants to the fishery, thus cancelling out any benefits gained.

Lake Victoria

This volume contains three papers on Lake Victoria, one considering the lake as a whole, and two drawn from Tanzania. This is the largest of Africa's lakes, and the second largest lake in the world. Covering 68,800 km^2, it is also Africa's largest inland fishery. Compared to its Great Lakes partners on the continent, Lake Victoria is remarkably shallow, averaging just 40 m. In Chapter 8 Geheb *et al.* set about considering the problems in the management of the fishery and then go on to propose a community-based management system to replace the present 'command and control' regime. They suggest a redefinition of state activities towards the provision of guidance, extension and other support services to community management units, and argue, possibly controversially, that the minimal impact of present regulations means that a system dominated by communities would certainly not make conditions any worse. Finally, the paper implies that corruption and misadministration are very much part of the management landscape in this part of the world, and suggest that it makes sense to develop a management system that can cope with this characteristic, rather than to pretend that it does not exist.

Wilson's paper from Tanzania first considers fishermen's perceptions of present management strategies and then how they react to suggestions for a future co-managerial system. (Chapter 9) he reveals how communities are evidently concerned about the need for a harmonised management strategy to ensure that communities all proceed at the same pace, and their worries that the size of the lake will rule out this possibility. This concern is presumably linked to their fears that there will be defaulters. They are also concerned that new institutions will be manipulated so that they benefit some more than they do others.

Wilson's work was carried out in the early 1990s, and Medard *et al.*'s paper (Chapter 10) demonstrates that he has judged the future well. This third Lake Victoria paper considers a deeply divided fishing community and how it attempts to use a new management institution to serve its own ends. Recalling the localised power struggles described by Hara *et al.*, Malasha and subsequently Krings and Sarch, Medard *et al.* (Chapter 11) argue that management must take into account that these conflicts are bound to occur, and rather than ignore and sideline them, management must face them and possibly even take them on board.

Lake Chad

Lake Chad's surface area is highly variable, from as little as 2,000 km^2 (in 1907) to 22,000 km^2 (in 1961), depending on rainfall within its basin. It is, on average, 3.9 m deep, and is shared between Nigeria, Chad and Niger (Vanden Bossche and Bernacsek, 1990). The lake is a vitally important wetland lying in the semi-arid Sahel corridor. It provides the basis for many thousands of livelihoods that depend on its seasonal fluctuations to renew fish stocks as well as to water farmland and rangeland. Despite the millions who rely on fish production from the lake, state management of the lake's fisheries resources is largely ineffective. In the early 1990s, immigrant fishermen from Mali and Hausaland introduced a new fishing technique involving traps, called '*dumba*' fishing. On the Nigerian shores of the lake, the new technique brought about various conflicts. Krings' and Sarch's case study (Chapter 11) outlines how communities in neighbouring areas of the Nigerian lakeshore addressed such conflicts and examines the institutions that were created to resolve them. Despite the promulgation of the 1992 Inland Fisheries Decree by the Federal Government of Nigeria, these institutions were created largely outside the sphere of government influence. Nevertheless, they have had some measure of success in conflict resolution and this chapter considers the insights that their evolution provides for the wider issue of inland fisheries management in Africa. The paper does not provide optimistic reading for those seeking to promote the co-management of fisheries between local communities and government fisheries departments. Rather, the case study demonstrates the importance of local elites in determining fishing rights, the extent to which their influence in non-fishing arenas enables them to manipulate interest groups within the fishery and the divergence between their objectives in doing so and those of the federal fisheries department (see also Hara *et al.* and Malasha in this volume for themes that to some extent echo these).

River Gambia

As this river enters Gambian territory, 680 kilometres from its source in the Fouta Djallon Highlands in Guinea, it flows along an east-west axis in the centre of the country dividing The Gambia into two halves. The river is rich in fishery resources with a variety of marine fish/shrimp species in the brackish water regime (about 250 km upstream) and freshwater species in the upper parts. Due to a lack of financial and human resources for stock assessment, research and monitoring, the fisheries resources potential of the river is not well known, but is considered to be under-exploited because of the low fishing effort. Large numbers of foreigners (from Senegal, Mali, Guinea and Guinea Bissau) exploiting the fishery have led to competition that is disadvantageous for nationals and threatens to jeopardise the state of the resource. NJie's and Mikkola's case study (Chapter 12) examines how Gambian fishermen have sought solutions to these problems in community-based management. Although not legally binding, these types of management form a basis for fisheries governance and call for the support and collaboration of the government in a co-management system of inland fisheries management.

The papers in this volume have in common their exploration of possible managerial solutions to Africa's fisheries problems. They emphasise the crucial role of diverse, heterogeneous and often disparate rural communities in the fisheries management discussion, and the need to seek fisheries solutions by addressing community problems. They often propose partnerships between the state and communities as a possible mechanism by which this may be achieved, but caution that these relationships will be complex and dynamic. It is against this template that future managerial strategies must operate.

2

Fisheries Co-Management in Malawi: Implementation Arrangements for Lakes Malombe, Chilwa and Chiuta[1]

Friday J. Njaya

The formulation of sound policies for the management of fisheries resources in Malawi faces a number of problems, such as:

(a) Complex species composition;
(b) Limited entry controls;
(c) Inadequate surveillance, monitoring and control of fishing activities;
(d) Lack of, or few, alternative income sources;
(e) Increased population growth;
(f) Climatic variability; and
(g) Environmental degradation.

Consequently, fish stocks such as the most important commercially viable cichlid (*Oreochromis* spp.: '*chambo*') have been declining, particularly in the south-east arm of Lake Malawi, the Upper Shire River and Lake Malombe.

In 1995, Lake Chilwa dried up. This resulted in the collapse of the commercially important stocks of endemic *Sarotherodon shiranus chilwae* ('*makumba*'), the cyprinid, *Barbus paludinosus*, and the catfish *Clarias gariepinus* ('*mlamba*'). On Lake Chiuta, *makumba* stocks declined due to the widespread use of fine meshed open water seine nets ('*nkacha*') by migratory fishers. These declines and the low compliance with regulations necessitated a refocusing of the fisheries management regime in Malawi. In 1993, the concept of 'co-management' was adopted as a fisheries management guiding principle, bringing about a shift in managerial perspectives from centralised 'top down' approaches, to a 'bottom up' approach in which resource users participate in formulating fisheries management policies. This resulted in the implementation of a pilot Participatory Fisheries Management Programme (PFMP) on Lake Malombe and the Upper Shire River to reverse the fisheries decline. The strategy employed to implement the PFMP involved the creation

[1] An earlier version of this paper was presented at the 1998 International Institute for Fisheries Economics and Trade (IIFET) Conference in Tromsø, Norway, July 8-11, 1998.

of a Community Liaison Unit (CLU) composed of fisheries extension staff, and Beach Village Committees (BVCs) representing the fishing communities.

Since the adoption of co-management on Lakes Malombe, Chilwa and Chiuta, improvements have occurred in various areas, including in the relationship between the Fisheries Department and user communities, in participatory licensing and enforcement activities and in the formulation and review of fisheries regulations by the fishers.

This paper examines the development and implementation of fisheries co-management on Lake Malombe and the Upper Shire, and Lakes Chilwa and Chiuta (Figure 2). Specifically, it examines various types of institutional arrangements and their historical background.

A general overview of the fisheries sector

The fisheries sector in Malawi plays a major role in providing the population with employment opportunities in fishing, processing or trading activities. A study conducted by ICLARM/GTZ (1991) indicated that, in addition to their direct subsistence importance, small-scale fisheries in Malawi provide the principal livelihood for a large number of its rural households. The sector offers direct employment to about 43,000 people and indirect employment to approximately 100,000 - 150,000 people according to the 1994 frame survey (Donda, 1996). The capture fisheries sector also contributes about 4% to the country's Gross Domestic Product (GDP). In terms of nutrition, fish forms a principal source of dietary protein in Malawi, providing over 70% (on a fresh weight basis) of the total animal protein consumed in the country.

The capture fisheries comprise three main sectors: the traditional sector which lands over 85% of annual fish production; the commercial sector, landing between 10-15% of the annual total catch; and, lastly, the flourishing aquarium trade.

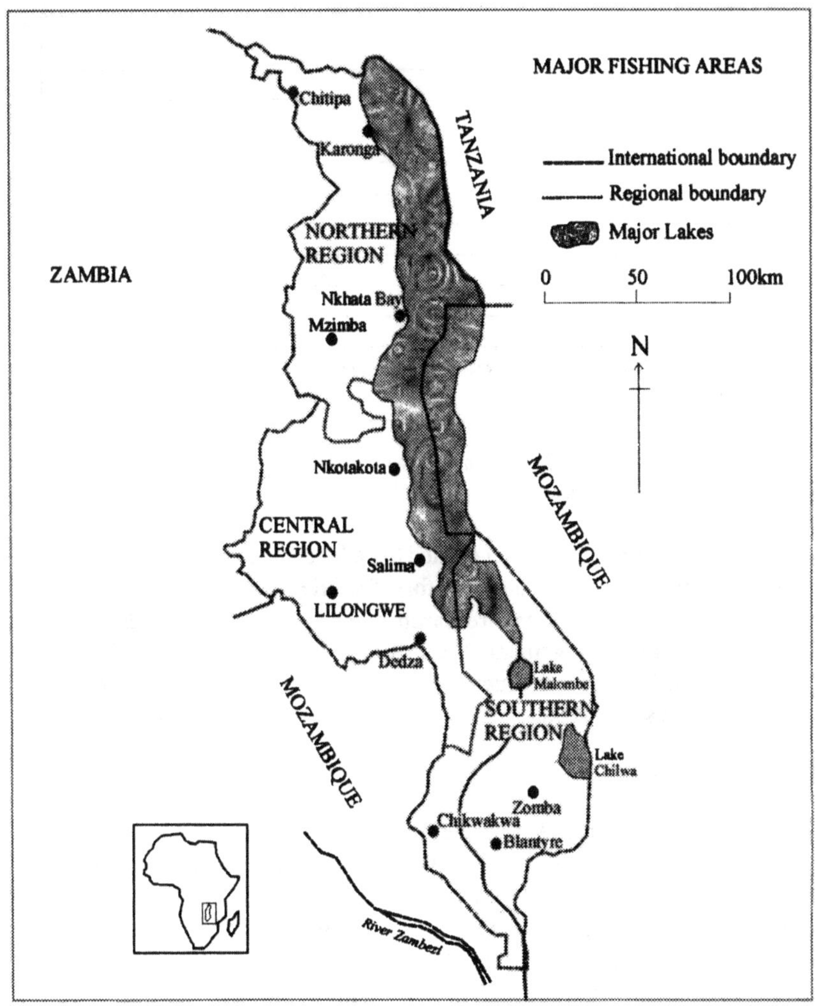

Figure 2: Lake Malawi

For a period of fifteen years (1982-1996), annual fish production in Malawi varied between an estimated 53,900 to 88,400 t. (Figure 3). The lowest production of 53,900 t was recorded in 1995 due, in part, to the drying up of Lake Chilwa. Malawi's fish production is dominated by catches from Lakes Malawi, Chilwa, Malombe and the Shire River system.

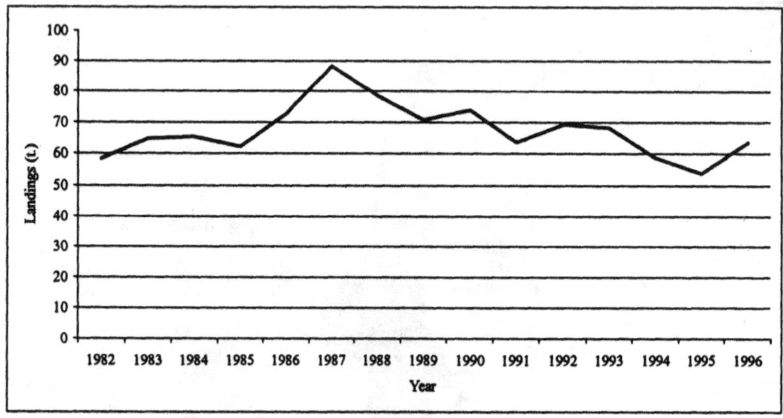

Figure 3: Estimated fish production in Malawi, 1982-1992.
Source: Fisheries Department.

Malawi's fisheries policy is being drafted to include various issues on fisheries extension with a focus on participatory fisheries management, training, aquaculture, enforcement and research components. The new Fisheries Conservation and Management Act was approved in October 1997 .with community participation, resource ownership and the empowerment of local communities as some of the additional sections to the previous Act.

The concept of 'co-management'

Fisheries managers pursue multiple goals. Rarely is the task of fisheries management defined in biological terms only. There are also social and economic concerns because the management of capture fisheries - as opposed to aquaculture - is aimed at people, and not only fish. What is 'rational' at societal level may be intolerable at local level; what is efficient from an economic perspective may be socially and culturally harmful; and what makes sense in biological terms may be unwise in social and economic terms. All things considered, a maximum sustainable yield is not necessarily an optimal yield (Jentoft and McCay, 1995). Such apparent contradictions may result in conflict amongst various user groups and policy objectives governing the utilisation of fisheries resources in Malawi, particularly in the artisanal sector.

Co-management, as defined by Sen and Nielsen (1996), is an arrangement where responsibility for resource management is shared between the government and user groups, and is one possible solution to the growing problems of resource over-exploitation. The concept focuses on the recognition that user groups have to be more actively involved in fisheries management if the regime is to be both effective and legitimate. In Malawi, the concept has recently gained legitimacy given that the effectiveness of centralised fisheries management system is being questioned, and as some fish stocks in certain, localised, fishing grounds continue to be either fully exploited or in crisis as was the case with Lake Malombe and the Upper Shire River in the early 1990s.

Community participation, as defined by Campbell and Townsley (1996), is the active, meaningful and influential involvement of individuals or groups in an activity. In the context of fisheries co-management, it means that individual fishers or fisher groups and other agencies through various types of structures are actively involved in the management of fisheries resources. If management is to succeed, fishers must support its efforts (Wilson *et al.*, 1994). Such support will be realised if communities have evidence that regulations are working in their best interests. Individuals who are required to engage in short-term sacrifice in order to obtain collective (or private) long-term benefits need to be assured that their sacrifice really will have a positive impact on the health of the resource.

However, as observed by Jentoft and McCay (1995), the degree of user group involvement may differ from one country to another. Correspondingly, organisation may also vary between the two extremes of government power, and fishers' power. In the first instance, fishers are at the receiver's end as fisheries management is an entirely top-down process. At the other extreme, fishers have full control, and organise and run their own management system, either through institutions that are basically informal or by means of a formal organisation, like a committee or co-operative.

The participatory fisheries management programme for Lake Malombe and the Upper Shire River

Lake Malombe and the Upper Shire River lie between latitude 14° 21' to 14° 45' South and longitudes 35° 10' to 35° 20' East, and are part of the Great Rift Valley system. Lake Malombe is on average 4 m deep, about 30 km long and 15 km wide. The Upper Shire River, about 12 km long, flows from the southern tip of Lake Malawi before widening to form Lake Malombe. The lake's surface area is approximately 390 km^2. The lake is further enriched by influent streams from its densely populated catchment area and by the recycling of nutrients in its sediments as a result of its shallowness. Lake Malombe is much more productive than Lake Malawi. In 1988, when the fishery was near its peak, the lake produced about 15,500 t of fish, approximately 17% of Malawi's total production.

Lake Malombe's fishery is dominated by *Oreochromis* spp. ('*chambo*') and *Haplochromis* spp. ('*kambuzi*') (FAO, 1993) (Figure 4). *Chambo* catches have declined rapidly in this fishery, from about 8,300 t in 1982 to less than 100 t. in 1994, representing a considerable income loss to fishers. The total catch value (Wilson, 1993), fell from about MK36 million in 1982 to about MK8.4 million in 1990. The 1997 annual frame survey results showed that an estimated 420 gear owners and 2,854 ancillary workers operated on Lake Malombe, while on the Upper Shire River 87 gear owners and 776 ancillary workers were recorded. Fishers used 475 planked boats and 112 dugout canoes (Fisheries Department, 1997).

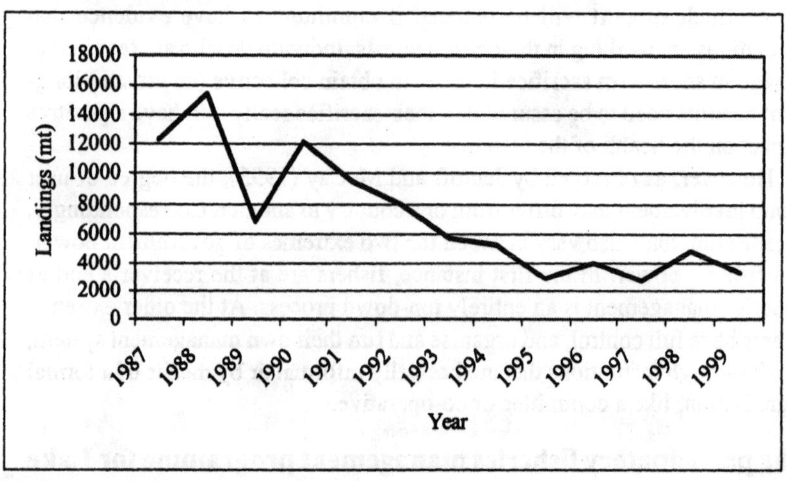

Figure 4: Estimated fish landings (t) from Lake Malombe and Upper Shire River (1987-1999). **Source:** Fisheries Department.

Gill nets are the most frequently used gear type along with *nkacha* nets, which account for approximately 99% of fish landings from Lake Malombe. Fishing is the most important income-earning activity for the fishers of Lake Malombe and the Upper Shire River. The problems associated with the management of artisanal fisheries in Malawi were first outlined by the FAO/UNDP-funded Chambo Research Project, implemented from 1988-1992 in the south east arm of Lake Malawi, the Upper Shire River and Lake Malombe.

In its concluding remarks, the Chambo Research Project's final report presented an alarming picture as *chambo* stocks in Lake Malombe collapsed due to the indiscriminate capture of juveniles by fine-meshed seine nets (UNDP/FAO, 1995). For this reason, the *kambuzi* was in danger of being over-fished.

The estimated annual fish production for Lake Malombe declined from about 12,000 t in the 1980s to nearly 5,000 t. in the early 1990s (Figure 4). The overall value of the fishery declined substantially, and with it, fisher-folk's incomes. In the Upper Shire River and south east arm of Lake Malawi all commercially important stocks were fully exploited. This meant no further increases in fishing could be sustained, and hence a management strategy was needed as a matter of urgency.

A number of management measures were introduced by the Fisheries Department to regulate entry into the fishery, protect breeding and juvenile fish through closed seasons and legal mesh sizes. It proved very difficult to enforce these regulations because of inadequate resources. In any case, a regulation which does not have the support of the fishing community, is often ineffective and typically not complied with unless it is enforced from beyond thecommunity.

The Chambo Research Project's findings paved the way for a new pilot fisheries management plan called the Participatory Fisheries Management Programme (PFMP), which was initiated by the Fisheries Department and focused on community participation or co-management. The Fisheries Department and the local community have been implementing the PFMP funding from the GTZ, UNDP/FAO, the ODA and the World Bank Fisheries Development Project since 1991.

The PFMP set up

In designing the PFMP, extension was considered a core component, followed by research and income-generating activity. With time, however, it was observed that enforcement by Beach Village Committees (BVCs) was not, in some cases, effective. The major co-ordinating linkages in the PFMP set-up are shown in Figure 5, and comprise the BVCs, the Ministry of Natural Resources and Environmental Affairs through the Community Liaison Unit (CLU) and the Fisheries Department. In some cases, there are local level linkages between the Area Executive Committee (AEC), composed of locally-based government department technical officers, and the District Executive Committee (DEC), composed of officers at higher levels of the district chain of command.

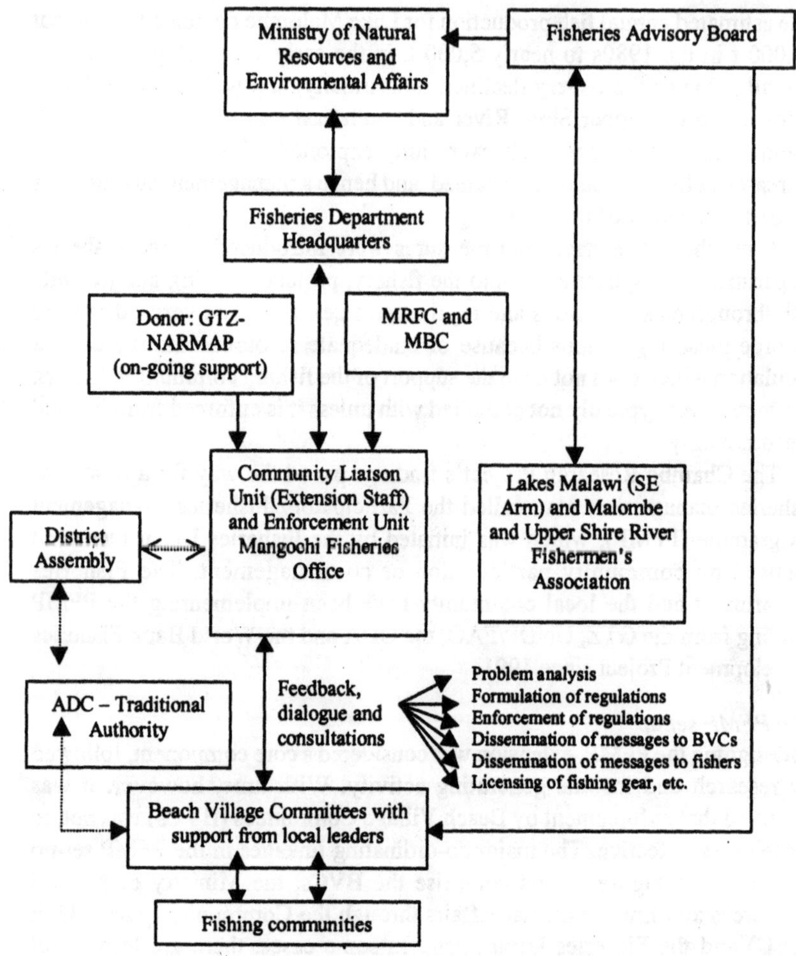

....	Co-ordination/linkage not fully functional
———	Co-ordination/linkage fully functional
MBC	Malawi Broadcasting Corporation
MRFC	Malawi Rural Finance Company for loans to women groups
NARMAP	National Aquatic Resources Management Project funded by GTZ

Figure 5: The current PFMP arrangement on Lake Malombe and the Upper Shire River (since 1998)

These linkages may include Fisheries, Forestry, Community Development, Agriculture and Tourism sectors, and provide technical solutions to problems affecting communities in various parts of the district through Village Development, Area Development and District Development Committee meetings. These structures were set up to ensure that demand-driven projects obtained support from donors and are now supported by the Decentralization Policy.

Lower down the chain of command, the newly formed Lake Malombe and Upper Shire River Fishermen's Association was created to act as an umbrella body for all BVCs and also to act as a link between the BVCs and the proposed Fisheries Board. The Fishers' Association and the BVCs collect fishing license fees from fishers, and have sometimes organised patrols in those areas where illegal fishing is known to occur.

Formation of the Beach Village Committees (BVCs)

The Beach Village Committees (BVCs) were intended to serve as the basis for a two-way channel of communication between user groups and the Fisheries Department. It was also hoped that they would progressively assume responsibility for the management of the fishery. To this effect, a total of 31 BVCs around Lake Malombe and the Upper Shire River were formed.

The BVCs are selected by village communities although in some instances the local leaders have appointed certain individuals for their personal gain. The BVC may be composed of members from the following groups:

(a) gear owners.
(b) Fishing crew members.
(c) Any active member of the village group.

It was, however, found that in most BVCs there are very few fishers and the majority do not have the crew members. This representation could be a very important factor in determining the performance of any BVC. The village heads are supposed to serve as advisors to the BVCs.

After several initial meetings facilitated by the Fisheries Department, the fishing communities identified the following roles for their BVCs:

(a) each BVC should control a named beach or beaches. The officers of the BVC and the members of the group it represents should be listed along with their gear.
(b) The BVC should control the admission of additional gear owners to the group.
(c) The BVC should control the use of each beach and thus limit access.
(d) The BVC should be prepared to expel members who do not comply with the agreed instructions, especially regarding closed seasons, gear specifications etc.

(e) The BVC should organise meetings to discuss the problems of the fishery and to reach decisions on how to solve them.

(f) The BVC should represent its members at higher levels e.g. through the recently formed Lake Malombe and Upper Shire Fishermen's Association which has the support of all the BVCs.

The Lake Malombe and Upper Shire River PFMP had a large number of external collaborators, including numerous international agencies (GTZ and the ODA, amongst others), and many internal, governmental collaborators (including various ministries, the Commercial Bank of Malawi, the Malawi Rural Finance Company, the Malawi Broadcasting Corporation and others) until 1999, when most of them had been phased out leaving only the GTZ to support the programme. The early phasing out of donor support has affected the performance of the programme resulting in the reduction of some of the management activities being carried out by the communities involved.

Performance assessment of the Lake Malombe and Upper Shire River PFMP
The UNDP/FAO (1994) reported that the benefits of the PFMP could not be assessed in years, but rather in decades. The report gave an example of the change of under-meshed *kambuzi* seines from less than 13 mm to 19 mm, which in the short run, would lead to smaller catches. In the long run, however, if regulations are obeyed, larger catches may be realised. Previous evaluation studies including Donda (2001) and Hara (2001) have indicated numerous shortfalls in the programme. Further assessment has been considered in Hara *et al.* (this volume), which has focused on conflicts between the local leaders and the BVCs, limited sanctions imposed by the BVCs on illegal fishers, failure by the Fisheries Department to meet its expected obligations such as revenue sharing and corruption tendencies among the BVCs and local leaders. However, the improved relationship between the Fisheries Department and the BVCs and the participatory formulation of regulations, licensing and enforcement activities by some of the BVCs have been noted as some positive effects of the co-management.

The Lake Chilwa participatory fisheries management programme

Lake Chilwa (Figure 2) lies 100 km south-east of Lake Malawi and at latitude 15° 30'S and longitude 35° 30'E, making it the most southerly of the major African lakes. It is a shallow, saline lake separated from Lake Chiuta by a sand bar some 20-25m high. The lake lies at a mean altitude of 622m above sea level and had an open water area of 678 km^2 in 1972. This is surrounded by a further 578 km^2 of swamps and marshes and 580 km^2 of seasonally inundated grasslands. These areas vary with the level of the lake in any year.

The total area of the ecosystem, including islands, is 1,850 km^2 (Kalk *et al*, 1979). Five main rivers flow into Lake Chilwa: the Domasi, the Likangala, the Thondwe, the Namadzi and the Phalombe. Between them, these rivers contribute 70% of total inflow to the lake (Kalk *et. al*, 1979).

The level of the lake fluctuates annually by 0.8-1 m, but larger fluctuations, of the order of 2-3 m, occur over periods of 6 and 12 years. When these oscillations coincide, as in 1967-68, the water level falls drastically and the lake may actually dry up. Another minor recession occurred in 1973, and two decades later, Malawi experienced a serious drought that caused Lake Chilwa to dry up. This event prompted the design of a Participatory Fisheries Management Programme (PFMP) to assist with the recovery of the decimated fish stocks.

The most abundant fish species in Lake Chilwa are *Sarotherodon shiranus chilwae*, *Barbus paludinosus*, ('*matemba*') and *Clarias gariepinus*. Other fish species found in the lake include the '*mphuta*' (*Gnathonemus* spp.) and '*nkhalala*' (*Alestes imberi*). In 1992, 2,496 fishers were operating on Lake Chilwa, while the frame survey of 2000 conducted by the BVCs showed that there were 2,689 gear owners and 6,570 crew members operating 1,868 dugout canoes, and 778 planked canoes. This shows that the number of operators on Lake Chilwa increased after its recession, which can be attributed to many reasons, one of which is that since the mid-1990s, not many Malawians travel to work in the South African mines.

The most common gears on Lake Chilwa are fish traps, gill nets, seine nets and long lines. In terms of craft, traditional dugout canoes dominate the fishery. From 1990 fish production ranged from about 7,000 - 11,000 t until 1996 (Figure 6), when the lake dried up resulting in the collapse of the fishery. Estimated fish production in 1999 was 12,500 t, almost the same catch figures as in the early 1990s before the recession.

The major problems observed as a result of the Lake Chilwa fisheries collapse included a reduction in employment opportunities, fish supply, water supply, and a decline in business activities and in the general welfare of the people. The problem was exacerbated by the limited migration of fishers out of the fishery. Following the 1967/68 Lake Chilwa recession, it took about a year for *Clarias gariepinus* to recover, while *Sarotherodon shiranus chilwae* and *matemba* stocks took two to three years to recover (Kalk *et al.*, 1979). To facilitate recovery after the 1995/96 recession, a number of strategies were proposed including artificial restocking by breeding *Sarotherodon shiranus chilwae* in ponds and then transferring them to the lake.

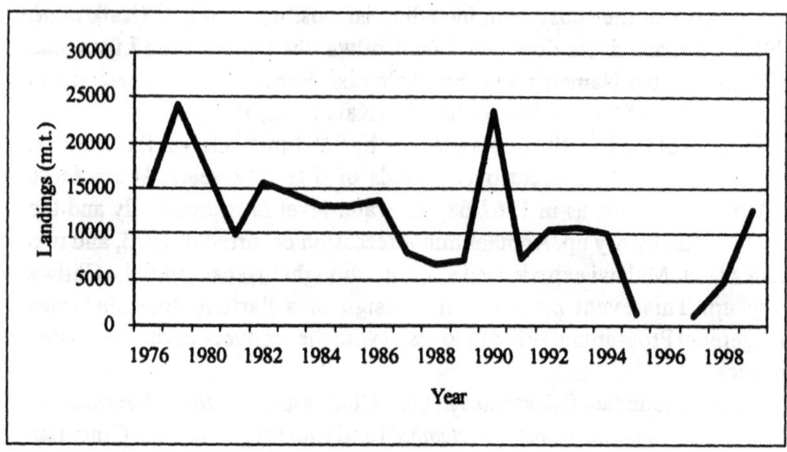

Figure 6: Estimated fish landings (t.) from Lake Chilwa (1976-1999).
Source: Fisheries Department

This proved problematic, however, due to limited funds, staff and technical skills. As a result, natural restocking was considered the only viable alternative.

Lake Chilwa's influent rivers play a major role in the ecosystem, and its fish stocks depend on the riverine and swampy ecosystems for breeding and the survival of juvenile fish. As such, the fish that existed in the lake were also present in the influent rivers, and it was therefore suggested that for natural restocking to occur, all fish stocks in reservoirs along the influent rivers should be conserved to repopulate Lake Chilwa after refilling. So as to make the task of enforcement easier, it was considered important to involve those communities settled along these rivers in the process.

Local leaders along the rivers were first sensitised to the importance of facilitating the formation of Beach Village Committees (BVCs). These BVCs were considered vehicles through which extension messages could reach the fishers, who could also participate in the enforcement of regulations, such as suspending seining on the rivers and Mpoto lagoon and banning the use of poisonous plants to fish with. It was in that way that the Lake Chilwa PFMP was established.

The Lake Chilwa PFMP set-up

The establishment of the Lake Chilwa PFMP is similar to Lake Malombe's in the sense that the Fisheries Department influenced the programme. It is, however, different from Lake Malombe PFMP in that the key co-management partners are local leaders, on the one hand, and the government on the other.

After the lake was flooded, local leaders continued to mobilise fishers and others to form BVCs. The Lake Chilwa and Mpoto Lagoon Fisheries Management Association was elected in 1999. However, because it comprises only local leaders, it has been criticised for not directly representing the interests of resource users (Lowore and Lowore, 1999), although it has been argued that the recognition of the traditional leaders' role is very important for sustainability of the co-management arrangement (Odote, pers. comm., 2001). In the meantime, there is a direct link between local leaders and the government (Figure 7). Local development structures (the ADC and DDC) are in place, although their linkage with the BVCs is not yet effective.

The Lake Chilwa PFMP has been largely implemented by the Fisheries Department with financial resources from donors like Germany Technical Cooperation (GTZ), especially between 1995-1998. Since 1999, there has been support in capacity building, message formulation and delivery and enforcement activities from the Danish International Development Agency (DANIDA). In 2000, the Community Partnerships for Sustainable Resource Management Systems (COMPASS), a non-governmental organisation funded by the United States Agency for International Development (USAID), supported the BVCs by providing three patrol boats for their use in the three districts sharing Lake Chilwa.

In March, 1997, all the local leaders around Lake Chilwa met and discussed a number of regulations. The regulations agreed upon were as follows:

(a) A closed season for active gears (seines) from 1 December to 31 March.
(b) A minimum gill net mesh size of 38 mm.
(c) Fishing from floating huts ('*zimbowera*') was prohibited because it was difficult to check the equipment being used on these.
(d) *Nkacha* seining is prohibited on Lake Chilwa.
(e) *Chambo* below 100 mm in length should not be landed.
(f) Letters signed by both the traditional leader and the BVC chairperson must be submitted by immigrant fishers on arrival at destination beaches.
(g) The use of poisonous plants ('*katupe*') for fishing on influents should be banned.

However, during a review workshop conducted in October 2000, local leaders and BVCs agreed to lift some bans, like the use of *zimbowera*, and they proposed a change to the closed season to be based on consultations and research.

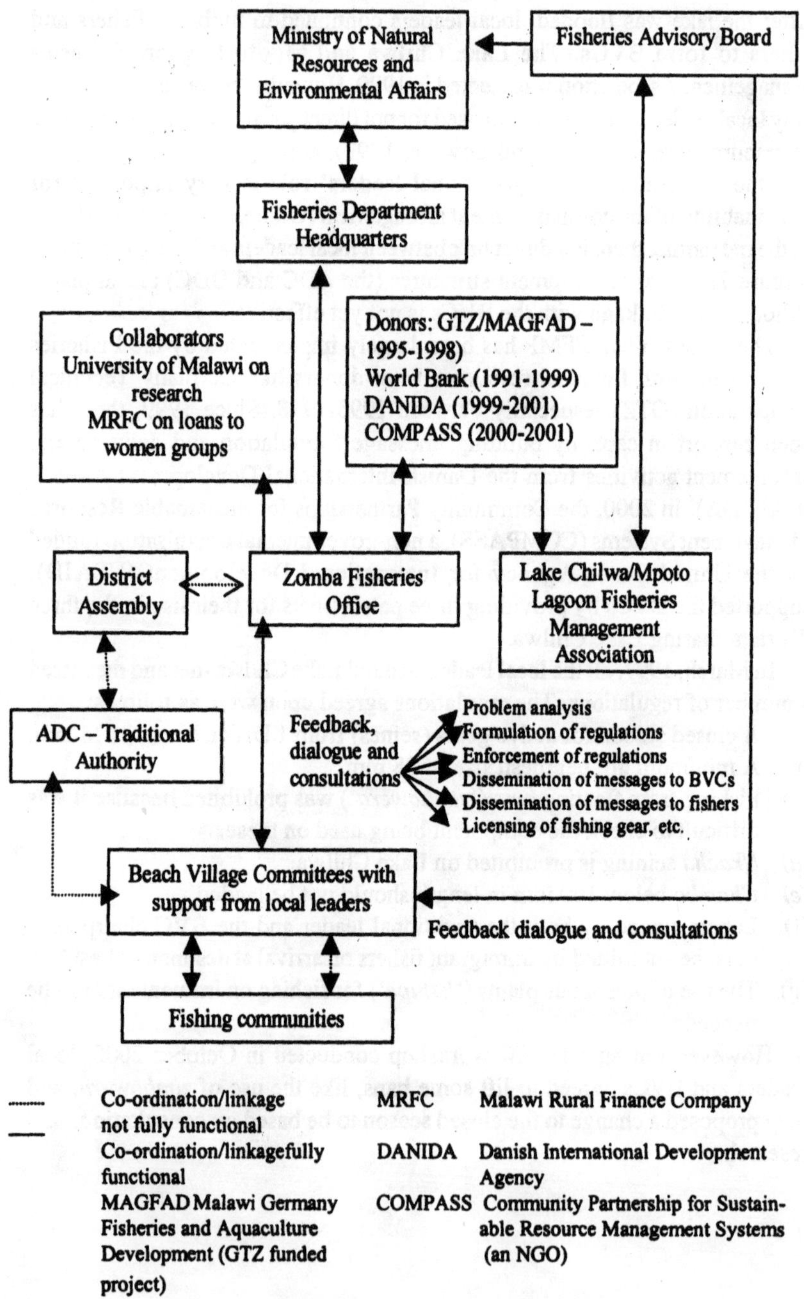

Figure 7: Arrangement of the PFMP on Lake Chilwa

Lake Chilwa filled to 'normal' level in 1997. Sampling experiments conducted in July, September and December 1996 and March 1997 indicated a steady increase in the weight of the samples (on a fresh weight basis) (Table 1).

	Fish species				
Month	Makumba (O. shiranus (chilwae)	Matemba (Barbus Paludinous)	Milamba (C. gariepinus	Others	Totals
April	186.63	151.79	59.98	3.48	401.88
May	157.11	303.98	76.23	5.37	542.69
June	105.37	170.24	130.62	18.23	424.46
July	144.02	271.79	79.57	6.32	501.70
August	137.17	282.06	88.20	2.26	509.69
September	112.63	184.26	62.90	1.00	360.79
October	162.08	269.45	90.37	0.20	522.10
November 1	24.19	193.08	58.73	9.55	385.55
December	38.66	16.68	95.79	6.41	157.54
Totals	1167.86	1843.33	742.39	52.82	3806.4

Table 1: Fish Production (t) in Lake Chilwa after refilling in 1997.
Source: Fisheries Department

The impact of community participation in the Lake Chilwa Recovery Programme

(a) Enforcement of the regulations was enhanced at the local community level because illegal fishing gear was confiscated from offenders.

(b) The fish stocks that survived the recession had been conserved by the resource users since 1996. In April 1997, local leaders around Lake Chilwa met to review the fishing suspension because some fish stocks were observed to have recovered through *ad hoc* sampling surveys conducted by the Fisheries Department.

(c) The participatory approach served as a basis for further management of the fishery as local leaders were in a position to set up measures that would result in the sustainable utilisation and management of recovered fish stocks on Lake Chilwa. Catch statistics indicate that fish stocks had almost fully recovered just two years after the lake refilled. Communities still participate in formulating regulations, and enforcement and review meetings are held at least once every year.

(d) Of particular importance was the involvement of the BVCs in conducting a frame survey in 2000, an activity that demands large financial resources for the Fisheries Department. It was only on Lakes Chilwa and Chiuta that the communities did the frame survey in 2000. This shows that co-management can be effective in reducing costs while achieving intended results.

The Lake Chiuta participatory fisheries management scheme

Lake Chiuta is located at an altitude of 620 m in the southern part of Malawi. It is a shallow lake with a mean depth of 5m and has a total surface area of about 200 km², of which 40 km² lie in Mozambique (FAO, 1994b). The southern part of Lake Chiuta is more or less permanently covered with emergent vegetation penetrable only by canoes. Its waters are clearer and less saline than those of Lake Chilwa. The lake is fed by a number of influents, and is sometimes connected by a swampy channel to Lake Amaramba, from which flows the Lujenda River, a major tributary of the Ruvuma River. The main influent rivers of Lake Chiuta include the Lifune, Chitundu and Mpili rivers. Lake Chiuta's associated marshes are separated from those of Lake Chilwa by an extended sand bar, some 20-25m higher than present lake levels (Kalk *et al.*, 1979).

Lake Chiuta's fishery is dominated by artisanal fishers using dug out or planked canoes. Since 1985, the estimated annual fish production for Lake Chiuta has fluctuated between 700 and 4,000 t (Figure 8).

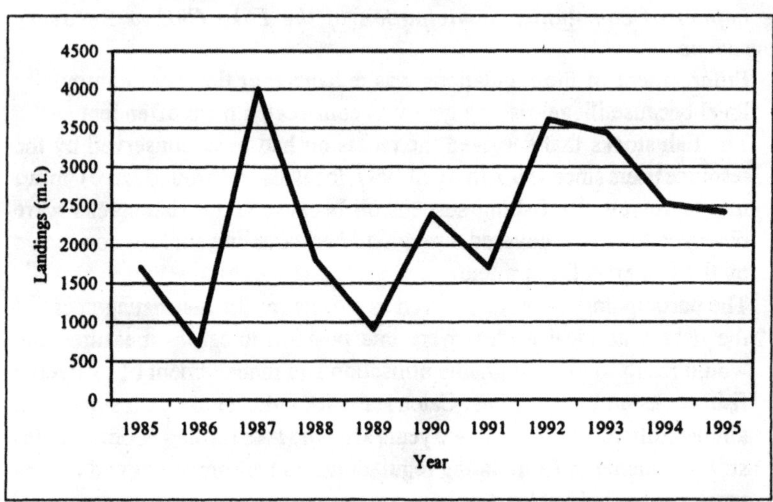

Figure 8: Estimated fish landings from Lake Chiuta (1985-1995).
Source: Fisheries Department

A frame survey conducted in 1990 showed that there were 462 fishers and 1,452 ancillary workers operating on the lake using 68 planked canoes and 282 dug out canoes. In a frame survey conducted by the BVCs in 2000, however, it was shown that 993 gear owners and only 38 crew members were operating on the lake, using one planked canoe and 564 dug out canoes. The reduced number of planked canoes and crew members is due to the eviction of the *nkacha* seine net fishers who used these. The indigenous fishers mainly use traps, gill nets, and long lines. Catches are dominated by *makumba* or *chambo*, *matemba* and *mlamba*.

Development of the Beach Village Committees

The initial arrangement of the Lake Chiuta fisheries management system was that of a community-based approach (Dissi and Njaya, 1995), with local leaders having powers to allow fishers to operate in the lake from their beaches. In the late 1980s and early 1990s, however, the number of immigrant *nkacha* operators on the lake increased. Local fishers objected to competition with these immigrants and claimed that the gear destroyed fish habitats and caught too many juvenile fish. The *nkacha* fishers landed larger catches than indigenous fishers, who operated mainly fish traps and gill nets. These catches drove down fish prices, adversely affecting trap and gill-net fishers. Because *nkacha* catches were large, however, immigrant fishers were still able to earn sizeable incomes, which indigenous fishers argued made them attractive to school-going girls, thereby affecting the girls' educational prospects. Local fishers formed a pressure group that sought the expulsion of the immigrants, and sought support from the Fisheries Department to have Lake Chiuta legally recognized in the Fisheries Act with a view to formulating and reviewing the regulations governing the exploitation its resources.

The indigenous fishers approached their local leaders and the Fisheries Department seeking to explain their position, and to obtain support. In some cases, however, local leaders were corrupt, bribed by *nkacha* seine owners to oppose these moves. Several meetings were held with the facilitation of the Fisheries Department without much success. Therefore, in May 1995, about 300 *nkacha* operators were overpowered in a confrontation and then evicted from the fishery.

With the facilitation of the Fisheries Department, Lake Chiuta's indigenous fishers elected nine BVCs on the Malawian side of the lake, and the Lake Chilwa Fishermen's Association (LCFA) was formed in 1996. In the following year the groups were trained in group dynamics, leadership roles and business management with funding from the GTZ MAGFAD Project.

The institutional set-up of Lake Chiuta's BVCs is different from those on Lakes Malombe and Chilwa (Figure 9) for two reasons: firstly, because the community influenced the co-management process and secondly, the arrangement is mainly based on banning a certain type of gear and not designed to enhance compliance with regulations.

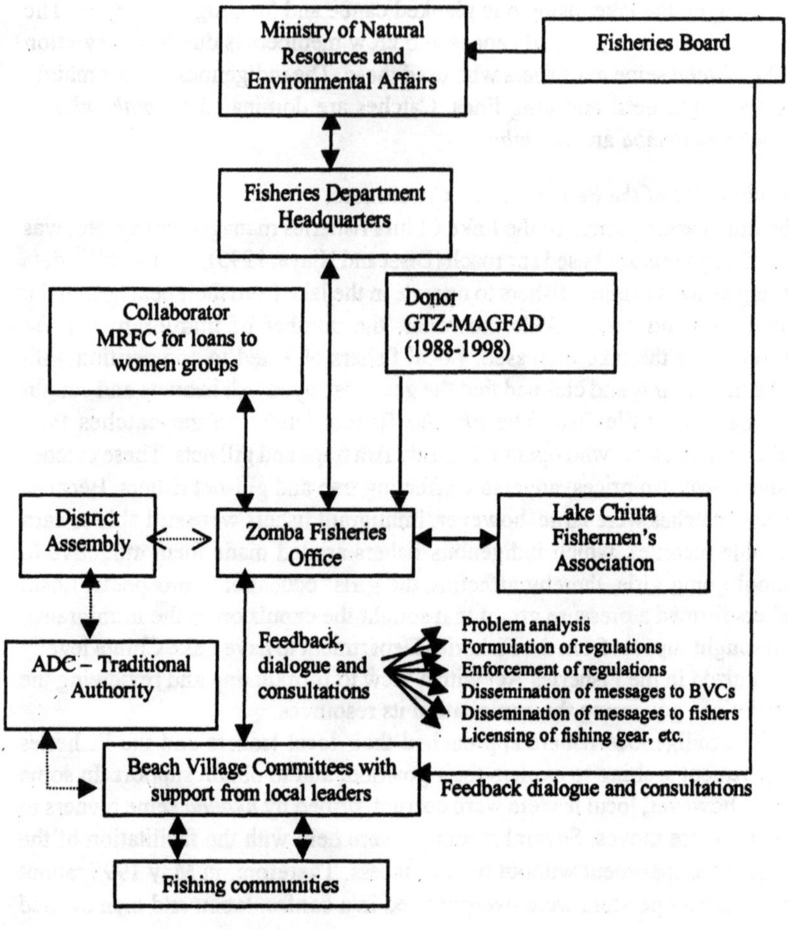

Ministry of Natural Resources and Environmental Affairs
Fisheries Board
Fisheries Department Headquarters
Collaborator MRFC for loans to women groups
Donor GTZ-MAGFAD (1988-1998)
District Assembly
Zomba Fisheries Office
Lake Chiuta Fishermen's Association
ADC – Traditional Authority
Feedback, dialogue and consultations
Problem analysis
Formulation of regulations
Enforcement of regulations
Dissemination of messages to BVCs
Dissemination of messages to fishers
Licensing of fishing gear, etc.
Beach Village Committees with support from local leaders
Feedback dialogue and consultations
Fishing communities

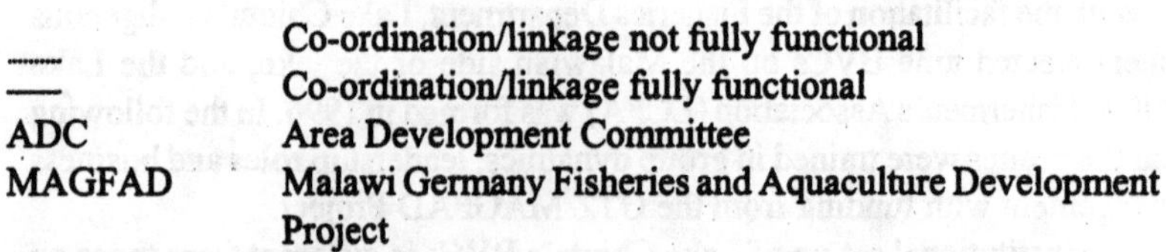

....	Co-ordination/linkage not fully functional
———	Co-ordination/linkage fully functional
ADC	Area Development Committee
MAGFAD	Malawi Germany Fisheries and Aquaculture Development Project

Figure 9: The Arrangement of the PFMP on Lake Chiuta

The regulations established by the Lake Chiuta BVCs were as follows:

(a) *Nkacha* seining is banned.
(b) The minimum mesh size for all gill nets should be 38 mm.
(c) Social conflicts should be settled by the BVCs with the support of traditional leaders, and with Fisheries Department staff as observers.
(d) The minimum takeable size of *chambo* or *makumba* is 100 mm (standard length).
(e) Beach seines are not allowed on Lake Chiuta and hence, there is no closed season for the lake.

Several benefits can be attributed to the eviction of the *nkacha* fishers from Lake Chiuta, including improved catches from gill nets, long lines and fish traps as reported by the BVCs (Njaya *et al*., 1999). In addition, due to the selectivity of gill nets, the average size of *chambo* or *makumba* landed has increased. With the implementation of the Lake Chiuta PFMP, relations between resource users and the Fisheries Department have improved. Fisheries regulations proposed by the fishers were approved by the Malawi government, and are reviewed and enforced by the BVCs and local leaders.

Some of the major problems encountered under the scheme include the following:

(a) Shared resource problems: as indicated earlier, Lake Chiuta is shared by Malawi (60%) and Mozambique (40%). On the Mozambican coastline, local leaders allow the use of *nkacha* seines. Conflict arises when Malawian BVCs then seize the *nkacha* seines and impose fines on their users. A joint management strategy is being worked out in an attempt to reduce these social conflicts.
(b) Reported cases of corruption by some local leaders who encourage the seine net owners to operate in the lake.

Considering that the Lake Chiuta fishery is a small-scale fishery, situated in a remote area with poorly developed infrastructure, it is important to ensure its sound management so that protein supplies and incomes are sustained for local fishing communities. Fisheries co-management is a suitable managerial strategy in this respect, because it offers an opportunity for fisheries regulations formulated by user communities to be perceived as legitimate, and therefore more likely to be complied with.

Summary and conclusions

The main features and managerial characteristics of the fisheries discussed in this paper are summarised in Table 2. Here, the main tenets determining the success or failure of the BVCs are considered.

Feature/Issue	Lake Malombe	Lake Chilwa	Lake Chiuta
Estimated size	390 km² (L. Malombe and 15 km of Upper Shire	About 1,850 km² including swamps and open water River).	200 km²
When co-management started	1993	1995	1995
Major results	Improved relationship between resource users and Fisheries Department Enforcement Unit	Recovery of collapsed fish stocks by conserving stocks fishers.for natural restocking of the lake after refilling.	Eviction of *nkacha* seine fishers.
Source of influence	Fisheries Department	Fisheries Department	Fishing community.
Key partners	BVCs with local leaders advising; and Fisheries Department.	Local leaders with elected BVCs and Fisheries Department.	BVCs with local leaders advising; and Fisheries Department.
Number of BVCs	31	42	9
How the co-management started	Lessons from other countries and meeting donors' conditions of support.	Application of local development self-help initiatives.	Lessons drawn from Lake Malombe PFMP.
Number of fishers (gear owners and crew) in 2000	No frame survey	8,656	1,031
Type of activities in which the BVCs have been involved.	Licensing, issuing of transfer letters, patrols.	Frame surveys, patrols, message delivery, issuing of transfer letters.	Frame surveys, patrols, message delivery, issuing of transfer letters.

Table 2: Major differences in issues or features among Lakes Malombe, Chilwa and Chiuta.

(a) *Size of Ecosystems*: the size of the water body on which co-management is being implemented is an important consideration. It would appear that the smaller the water body, the more likely co-management is to work. This prompts the question as to whether or not co-management will work on a larger water body, such as Lake Malawi.

(b) *Funding for co-management*: a lot of resources have been provided by various donors for the implementation of the Lake Malombe PFMP but not on Lakes Chilwa and Chiuta. The success and sustainability of such co-managerial approaches may, therefore, be a function of continued government or donor funding.

(c) *Composition of BVCs and fishers' associations*: the question as to whether or not the local level institutions should comprise fishers alone has prompted two schools of thought: on the one hand, local leaders are viewed as a necessary catalyst for co-management, while, on the other hand, they are seen as an impediment because they are vulnerable to corruption. Further investigation is needed. It may well be that what happens and works well in one area may not be applicable in other areas.

(d) *Effect of variability of an ecosystem on the resilience of co-management*: this is an issue mainly with respect to climatic variability that seasonally or periodically affects water levels in shallow and enclosed ecosystems like Lake Chilwa. It was observed that seasonal water level changes influence the migratory patterns of fishers. The question that should be considered is whether or not the cohesion of BVC structures may be guaranteed when these fishers migrate to other areas, or when the lake again recedes. This will need further research that seeks to understand how the composition of the BVCs affects their sustainability, as well as the contribution of local leaders.

(e) *Effect of co-management on socio-economic profiles and stock levels*: data on the socio-economic profile of fishing communities, and on fish catches and stock levels are important for monitoring purposes. Catch levels may be affected by factors such as the environment, and it is therefore difficult to gauge whether or not co-management has affected them. Co-management may not result in higher catches in the short-term as some of the measures being proposed may lead to reduced effort or landings. For example, a regulation on using larger mesh sizes could affect the size of fish landings. There is a need to estimate the biomass of the three lakes and make recommendations regarding exploitation patterns in consultation with resource users.

(f) It would be prudent to consider the various attributes that may have an impact on stock levels. However, the success of co-management may also be evaluated by looking at the partnership between the Fisheries

Department and local level institutions, the formation of BVC structures and the capacity of the community to formulate and enforce regulations.

(g) *Policy and operational issues*: the need to finalise and operationalise fisheries policy cannot be overstated. It is also important to look to other environmental policies, such as decentralisation and forestry, which complement the Fisheries Policy, so as to seek the effective implementation of co-management programmes. The political will and the readiness to devolve authority over resource management to communities are prerequisites in co-managerial arrangements.

The on-going co-management initiative in Malawi may offer alternative fisheries management options. Their success, however, may depend on the driving force for the scheme, which should be initiated by the resource users and not imposed on them by outsiders. It has been observed that it is cost-effective (e.g. by conducting frame surveys, enforcement patrols, extension and licensing activities), although no data is available to support this statement. In some cases, the BVCs serve as vehicles through which extension messages and community needs may be assessed within a limited time. The legitimacy of fisheries regulations - especially for Lakes Chilwa and Chiuta - has been enhanced. The process of consultation and collaboration in decision-making, however, sometimes delays certain activities and programmes. This may be addressed through good planning, co-ordination and the availability of all necessary resources. Participatory fisheries management needs to be supported in all aspects, such as through political will, legal issues, capacity, income-generating activities, gender considerations, research programmes and participatory extension skills. More lessons, however, are yet to be learnt for the future improvement of the on-going PFMPs and other programmes.

In terms of application, it may be the concept and principles of fisheries co-management that need to be applied and not an already designed arrangement being implemented elsewhere. In addition, co-management is not a solution to all problems associated with fisheries resource management. It is important that more studies are conducted to understand each partner's expectations, the socio-economic situation of the communities, the biological status of fish stocks, cultural aspects and the technological development of such fisheries.

Transboundary issues are important in the co-management arrangements for Lakes Chilwa and Chiuta as these ecosystems are shared between Malawi and Mozambique. If not checked, these issues may negatively affect the resilience of co-management programmes.

3

Lessons from Malawi's Experience with Fisheries Co-Management Initiatives

Mafaniso Hara, Steven Donda and Friday J. Njaya

The last two decades have witnessed a paradigmatic shift in fisheries management from state-centric strategies to co-managerial ones. Two such arrangements were introduced in Malawi on Lakes Malombe and Chiuta in 1993 and 1995 respectively. While the introduction of co-management was in response to a particular crisis on each of the lakes, the government motivated and facilitated the partnership in the former while fishers initiated it the latter[2]. Because of these different backgrounds, the two cases offer an interesting perspective on how the histories of the way in which such partnerships are formulated can be important determinants of the institutional arrangement, patterns of interaction and the outcomes of such managerial regimes.

Although the definition of co-management remains widely debated, most scholars and researchers agree that co-management is a form of institutional and organisational arrangement between government and user groups for the effective management of a defined resource (Berkes, 1997). This broad definition of co-management will be used for the purposes of this paper. The general functions of co-management can be identified as: the encouragement of partnerships; the provision of incentives for the sustainable use of resources at the local level; and the sharing of power and responsibility for conservation. As a management strategy, co-management is a compromise between government concerns for the conservation and efficient utilisation of the resource on the one hand, and resource users' concern for equal opportunities, self-determination and self-control on the other. Co-management is based on two assumptions: that local people must have a stake in conservation and management and that partnership between government agencies and resource users is essential for sustainable exploitation. Co-management advocates a

[2] The difference in importance attached to the two lakes can be seen in that more research and management attention had been given to Malombe than to Chiuta by the Fisheries Department. In its 1971 annual report, the Fisheries Department indicated that Lake Chiuta was not a priority in its management and conservation mandate because it was sufficiently remote and relatively unimportant.

shift away from autocratic and paternalistic modes of management to approaches that rely on the joint efforts of government agencies and users. The basic concern is the reshaping of state interventions so as to institutionalise collaboration between administration and resource users and end unproductive situations where one was pitted against the other as antagonistic actors (Baland and Platteau, 1996).

In the Western approach, three arguments have been used to justify the increasing adoption of co-management: that concerned interests should be heard; that information from users will result in the improvement of management decisions; and that co-management would improve the legitimacy of the management system, thereby reducing 'transaction costs' (Hanna, 1995; Hersoug and Rånes, 1997). In developing countries, community participation is increasingly being introduced as a condition for development aid in the resource management sectors as part of the democratisation process (Hara, 2001). Apart from the preceding arguments, there has been growing recognition - among managers, scientists and politicians - that no management scheme will work unless it enjoys the support of those whose behaviour it is intended to affect (Hersoug and Rånes, 1997). Evidence shows that if fishers willingly accept the regulations as appropriate and consistent with their existing values, the regulatory agency and the scheme will gain legitimacy with the fishers (Kuperan and Abdullah, 1994). Such regulatory schemes are associated with much reduced non-compliance with regulations. It is in this area of securing legitimacy that co-management shows promise of being a better resource management approach for fisheries than sole state management. As Hersoug and Rånes (1997) have stated, while the problems of lack of co-management and fishers' participation have been widely documented, it remains a challenge to demonstrate what kind of co-management has been successful and under what conditions. This paper is a contribution towards this challenge, particularly in the context of sub-Saharan Africa.

This study draws upon the Institutional Development Analysis (IDA) framework (Oakerson, 1992) as adapted for application in the Coastal Resources Co-management Project (Sen and Nielsen, 1996; ICLARM/IFM, 1998). This approach enables comparisons to be made between case studies, country research and co-management models (Sen and Nielsen, 1996; ICLARM/IFM, 1998).

Reasons for seeking an alternative regime

On both Lake Malombe and Lake Chiuta, the decline in catches was one of the main reasons for seeking an alternative management regime. Thus, while the overall objective of fisheries management in Malawi is stated as being the maximisation of sustainable yields from fish stocks that can be economically exploited from natural and man-made waters (Government of Malawi, 1987), the specific objective for co-management in the short term was to halt catch declines and put the lakes' fisheries on a course for recovery. While there seems to have been a reversal in negative catch trends and a stabilisation of catches on Lake Chiuta, this has not been the case on Lake Malombe[3]. Four possible reasons can be suggested for the lack of noticeable positive change in catch trends on Lake Malombe: (a) that the regulations put in place to stop and reverse the decline do not go far enough to address the problem of over-fishing (both recruitment and growth); (b) that users have ignored the new regulations and are still using under-sized meshed nets and violating the closed season; (c) that it took time for the programme to start functioning fully and thus start showing its impact; and (d) that other factors, such as ecological or environmental change, could be responsible for continued low productivity.

According to Banda (1996), experimental fishing trials done on Lake Malombe over a three year period (1994-96) showed that 40% of catches from legal ¾ inch (19 mm) mesh size *nkacha* nets were juvenile. Banda also pointed out that a gear selectivity test between a ¾ inch mesh size net and another of less than ¾ inch showed that the former caught 54% less fish than the latter. Implementation of the ¾ inch mesh size regulation therefore had serious socio-economic implications for the fishers. There was a strong possibility that even fishers who had changed their nets to ¾ inch were still lining their bunts with mosquito netting to keep their catches healthy. In any case, 40% of the fishers had not changed to the agreed ¾ inch mesh size minimum by 1995 (Mtika, 1996; Jumpha, 1996). Because of various project-specific implementation problems, the Lake Malombe/Upper Shire River Participatory Fisheries Management Programme did not fully take off in 1993. '*Kambuzi*' (*Lethrinops* spp.) is a fast growing species, and average fish sizes increase dramatically soon after a properly observed closed season. Therefore, signs of recovery would have become apparent within a year or two if there were any. Thus, the late start of programme activities could be discounted as an important factor for the lack of noticeable positive change in catch trends.

The FAO (1993) have suggested that rainfall run-off contributes significantly to the productivity of Lake Malombe. Apart from the drought conditions of 1994/95, rains in the area were largely normal during the 1990s. Thus, while the negative impact of environmental or ecological trends cannot completely be discounted in our understanding of catch declines, these seem unlikely given the good rains during the decade. If the inadequacies of the regulations, along with possible environmental or ecological factors, are put

---

[3] Refer to Chapter 2 for production figures from the fisheries considered in this chapter.

aside we are left with the continuing non-observance of the regulations as the most probable reason for the lack of positive progress in the recovery of Lake Malombe's fishery.

Apart from the expulsion of the *nkacha*, the reason for the stabilisation of catches on Lake Chiuta is said to be because fishers adhere to regulations. Adherence has been further enhanced under co-management (Njaya *et. al.*, 1999).

The question to pose therefore is why the shift to a co-management regime has had a positive impact on Lake Chiuta while this has not been the case on Lake Malombe? Evidently, on Lake Malombe the regime has not, as yet, resulted in any really noticeable changes in fishers' behaviour that could be translated into sustainable patterns of exploitation. Fishers have continued to use illegal nets and to disregard other regulations, such as the closed season (Banda, 1996). Furthermore, measures that were intended to introduce limited access in order to control and reduce capacity are viewed as unwelcome by most fishers. This is in contrast to Lake Chiuta, where co-management seems to have resulted in a greater adherence to regulations and continued access controls on 'destructive' fishing gears.

Reasons for present outcomes and lessons that can be drawn

While many factors may have contributed to the present contrasting outcomes on the two lakes, the reasons for the lack of genuine positive impact of the new regime on Lake Malombe can be explained around a cluster of factors, the main ones being:

(a) The role of government in the partnership and the organisational set-up.
(b) Power and authority struggles between the new management groups, the Beach Village Committees (BVCs), and the existing traditional authority (village headmen).
(c) The extent to which vested interests in the fishery participate in the management regime.
(d) Issues of legitimacy and incentives for participation.
(e) The effects of limited short-term external (donor) funding to the programme.
(f) Ineffective and ambiguous enabling fisheries legislation (Fisheries Department, 1997).
(g) The effects of prevailing socio-economic conditions on operational decisions and resulting dilemmas around the implementation of limited access.

These factors have greatly influenced the levels of compliance to regulations and, in the end, are likely to be crucial for the sustainability of the new regimes. We elaborate on each of these below.

The role of the Fisheries Department

In the Lake Malombe/Upper Shire River Programme, the Fisheries Department's role as facilitator and funder (mostly through donor projects) had a great influence on co-managerial objectives. In addition, fishers allege that the Department influenced the composition of the Beach Village Committees[4] (BVCs). Due to this, the BVCs were seen to represent the Fisheries Department, and most BVCs felt that they derived their powers and authority from the Fisheries Department. Because the BVCs assumed what are still viewed as Fisheries Department duties, such as enforcement and licensing, this perception was further enhanced, resulting in the alienation of BVCs from their constituency, and greatly affecting their effectiveness and authority within fishing communities.

In contrast, the user-based arrangement on Lake Chiuta was largely mobilised by the fishers, with the objective of getting support and approval for the expulsion of migrant *nkacha* fishers ('*nkacha*' seines are rectangular inshore seine nets used for catching the off-shore *kambuzi*). Because the impetus came from the fishers themselves, the fishers saw the objectives and initiative as having been largely determined by themselves. Since the fishers conducted the elections on their own without the facilitation of the Fisheries Department, there is much stronger feeling that the BVCs are their own creation.

Thus, in the fisher communities, the legitimacy of the BVCs is related to their perceived source of authority and also the extent to which they are seen as representing their members. In addition, the source of initiative for the partnership is crucial for the sense of ownership of the process and the regime. In this context, there is a much greater sense of ownership of the BVCs and their functions on Lake Chiuta than on Lake Malombe. This has also meant that in the case of Lake Malombe, the government has had to continue driving the process. When such support is linked to short-term (donor) project funding, the effects of reduced support after projects phase out are proving increasingly negative on the future of the BVCs and the regime as a whole, as will been show below.

Struggles for power and authority

By custom and historical tradition, village headmen derive economic privileges from the fishery through their positions. This can also be traced to the colonial era whereby under the 'indirect rule' policy (Mamdani, 1996), village headmen were allowed to collect monetary or material taxes. While village headmen

---

4 BVCs are committees that had been instituted through a democratic process as vehicles for the participation of fishing communities.

are no longer allowed to collect taxes in the form of money, the system of coerced material contributions to village headmen remains widely practised and enforced by the headmen themselves and their elders. In this context, all fishers landing their fish in a given village headman's area are supposed to give him a determined amount of fish, locally called '*mawe*', every week as an honorarium. When a migrant fisher comes into a village, he has to seek permission from the village headman to stay. By custom, the immigrant has to pay something to the village headman as a token of thanks. He will also have to pay the weekly *mawe* once allowed to stay. In addition, it is important to note that, in a village setting, decision-making and authority have historically been dominated by the village headman (Mamdani, 1996). While issues may be discussed in an open forum, or by a council of elders, and a consensus reached, the village headman retains the ultimate authority for making the final decision. Village headmen can even issue authoritative decrees without consultation (Donda, 2001).

The BVCs were envisaged to be strong independent bodies that could eventually assume delegated management responsibilities from government. But it was not as if BVCs were occupying a power vacuum. Indeed, some of the roles and functions of the BVCs infringed on the powers, authority and economic privileges of village headmen. For example, in order to control access for outsiders under the co-management arrangement, incoming migrant fishers were supposed to seek permission from *both* the village headmen and BVCs instead of from the village headmen alone. This directly infringed upon the powers and privileges of village headmen.

As a result, the creation of BVCs has resulted in contests for power and authority with the village headmen, especially in the Lake Malombe/Upper Shire River where the organisational model gave village headmen presumed honorary positions on the BVCs as ex-officio members. In this position within the BVC, the village headman became *de facto* leader of the BVC, even though he was not the chairman. In any case, village headmen in Lake Malombe/Upper Shire River have been prone to ignoring the authority of BVCs, since by historical tradition and custom, they hold ultimate authority. Where the BVCs have resisted being taken over by village headmen and have established some semblance of independent authority, fishers (on Lake Malombe/Upper Shire River) have often been confronted with dual authority as village headmen have continued to exercise their traditional authority on their own (Donda, 2001). In some cases, gear owners saw the new arrangement as a chance to cut out the practice of giving *mawe* and encouraged BVCs to challenge village headmen on this matter. The challenge on their powers and privileges left strong village headmen with no option but to try and curb the powers of the BVCs. It is in this context that village headmen, such as Kadewere from East

Malombe, used their authoritative positions to disband the elected BVCs and replaced them with BVCs that they could control or manipulate (Hara, 1996). In such cases, the fishing communities felt that BVCs largely represented the interests of the village headmen, which might not coincide with their own.

It is interesting to note that village headmen have been largely kept out of the BVCs on Lake Chiuta. Since the fishers elected the BVCs on their own without the facilitation of the Fisheries Department, they were able to ignore the Lake Malombe model that incorporated village headmen as ex-officio members. To a large extent, this was also due to the fact that the village headmen had alienated themselves from the fishers following allegations of corruption and collusion with *nkacha* fishers. Village headmen were said to have given no support to local fishers in their quest to expel migrant *nkacha* fishers because, allegedly, they had been bribed. Fishers had to solicit the support of government officials (the Fisheries Department, the District Commissioner and the Police) on their own in order to effect the expulsions. Even then, the local fishers had to forcibly evict *nkacha* fishers who refused to leave. This was because some village headmen had tacitly continued to give them permission to stay even after they had been told by the government to leave. Because of the negative role village headmen had played over the issue of evicting migrant *nkacha* fishers, local fishers felt it unacceptable to have village headmen as members of BVCs. The fishers have argued that village headmen should limit themselves to the administration of land issues and leave the fisheries to fishers (Njaya *et. al.*, 1999). Fishers, with the apparent connivance of the BVCs, have increasingly refused to give *mawe* to their village headmen (Njaya *et. al.*, 1999). The exclusion of the headmen from the BVCs, and the growing reluctance of fishers to give *mawe*, has increased the tension between the BVCs and village headmen. The worry for the future is that while BVCs have so far successfully managed to keep *nkacha* fishers out of Lake Chiuta (on the Malawi side), it remains to be seen whether they can continue to do so without the blessing and involvement of village headmen and without resolving the growing antagonism between these two power bases.

While there is a need to have reasonably independent management bodies (in this case, BVCs) it would also appear necessary to ensure that these are in some way complementary and accommodating to the existing traditional authority and the powers of village headmen. It is likely that the traditional and customary system of justice under the Traditional Authorities will have to play a role in administering justice and sanctions in community-based management arrangements since village headmen preside over traditional courts and informal systems of justice in their areas of jurisdiction (Hara, 2001). Thus, village headmen cannot be completely excluded from a community-based regime if such a regime is to have the ability to apply sanctions locally. The Lake Malombe situation suggests that building new institutions around the existing traditional authorities could provide a possible solution to the struggles for power and authority. Since traditional authority institutions are not seen as incorruptible, however, this necessitates the formation of new organisations that should be reasonably independent from traditional authorities.

What can be deduced from the preceding sections is that one of the most critical aspects in the introduction of co-management arrangements is the tension created around two organisational aspects. The first problem concerns the struggle for authority and power between the (supposedly) democratically elected management bodies such as the BVCs, and the existing hereditary (undemocratic) traditional authorities. The second source of tension can be attributed to the source of initiative and drive for co-management, whether this has been top-down from government or bottom-up from the users themselves (Figure 10).

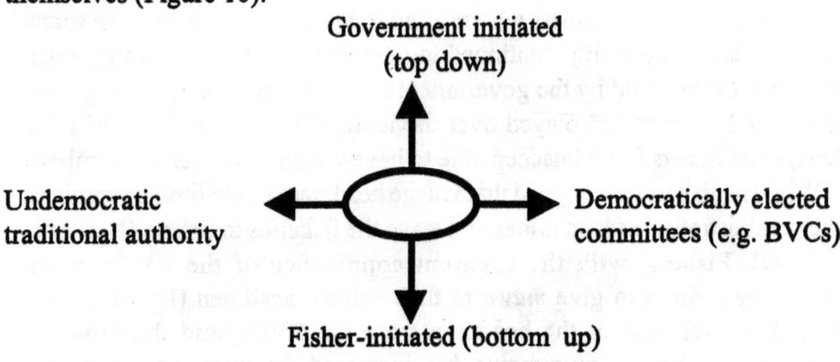

Figure 10: Two dimensions of tension relevant to introduction of co-management in Malawi (adapted from Hara, 2001).

The factors considered in Figure 10 need to gravitate towards the centre to produce effective and legitimate management bodies and institutional reforms. For the regime to be seen as being legitimate and representative, the fishing community must feel that they own the management process and its organs in terms of representation, the balance of initiatives and the election of their members. Too much influence from the Fisheries Department or village headmen results in a process where institutions are seen as alien to the stakeholders they are intended to represent. Thus, in the crafting of local management institutions sufficiently independent to command the respect and confidence of fishers, it is necessary that, firstly, village headmen are not rivalled or antagonised, and secondly, that these initiatives in no way contradict the aims and objectives of government. These are probably the most formidable but necessary tasks in the creation of co-management in Malawi's fisheries.

As a focus for user participation in the partnership, the BVC is located amongst three forces: the Fisheries Department, village headmen and the user community, each seemingly straining for control over, and to influence, the BVCs. In such a position, BVCs have to achieve a balancing act amongst the three in terms of the influence and derivation of power and authority (Figure 11). The government's main aim for the BVCs is that they should act as a vehicle for implementing and achieving its conservation objectives. At the same time, the BVCs must contend with the intolerance of village headmen.

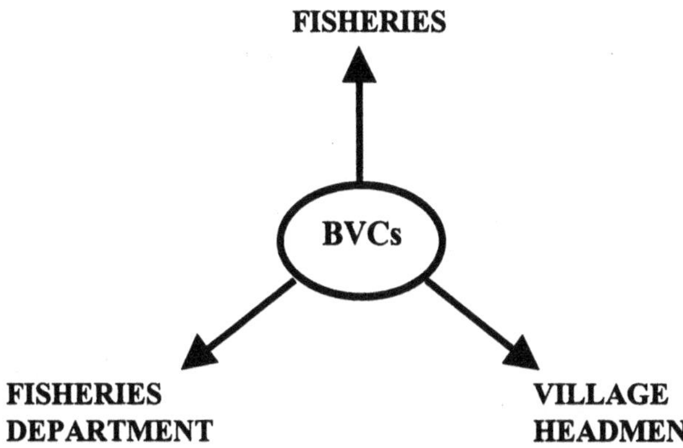

Figure 11: The three forces straining to influence the BVCs as a management institution (adapted from Hara, 2001).

While trying to keep the balance between government and village headmen, the BVCs must try to fulfil their obligations to represent the interests of their constituencies - the fishing communities - even if these might not be in line with the objectives of the Fisheries Department or the interests of village headmen. Pertaining to this, the BVCs on Lake Chiuta are more influenced by their members than those on Lake Malombe. As a result, users on Lake Chiuta see the BVCs as theirs, acting on their behalf and in their interests. This is in contrast to the situation on Lake Malombe, where most BVCs are seen as being influenced by either the Fisheries Department or village headmen. It would appear, therefore, that for an effective, locally-based, regime, the influence of users on management and representative bodies should be greater than those of the Fisheries Department and/or village headmen. Thus, how the powers and authority of the BVCs are related to those of the Fisheries Department and village headmen is one of the main challenges to co-management in Malawi's fisheries. Improving the extent to which the BVCs represent their constituencies also remains an important consideration.

Organisational structure, participation of vested interests and incentives for participation

The problem of the active participation of vested interests in fisheries management is crucial. This relates to gear ownership and the organisational structure of the industry on the two lakes. On Lake Malombe/Upper Shire River, the majority of gear owners employ crews to fish on their behalf. In this, operational decisions out on the lake are taken by the crew members. In terms of the Fisheries Act, it is the gear owner, not the crew member, who is legally responsible for any infringement of regulations. In addition, crew members' security of tenure in a fishing unit depends on their performance. As a result, there is great pressure on crew members to increase catches in any way possible. For the gear owners, keeping a good and productive crew is vital. In this sense, gear owners tend to avoid interfering with operational decisions that might negatively affect the size and value of the catch and, therefore, the benefits that the crew members get. In general, the sharing systems and the lack of long-term tenure within fishing units make crewmembers prone to operational decisions based on short-term economic maximisation strategies, and so encourage illegal fishing activities. In addition, because crew members lack legal responsibility for their actions, few deterrents exist to infringing regulations. The situation on Lake Malombe is in contrast to that on Lake Chiuta, where most gear owners are also active fishers. This means that gear owners are responsible for most of the operational decisions (and illegal activities, if any) that might occur out on the fishing grounds.

On Lake Malombe, fishers (gear owners and crew members) comprised only 30% of BVC members compared to 80% on Lake Chiuta in 1999 (Donda, 2001). Most of the former 30% were gear owners rather than crew members. As a result, most fishers in Lake Malombe shun or boycott meetings called by the BVCs and, to a large extent, ignore the resolutions that are passed by the BVCs. They complain that the BVCs take decisions on fishing issues that they have little knowledge about. Some fishers have gone as far as saying that they feel that the BVCs take decisions that are meant to punish fishers. For example, most fishers said that the BVCs did not consult them about the proposal to change the closed season (from January-March to November-December) taken in 1997 (Donda, 2001), and claimed that the BVCs did not have the mandate from the fishers to make or agree to such a proposal. Another effect of the high percentage of non-fishers on BVCs at Lake Malombe, is that members feel strongly that they should be paid for their services since they do not benefit directly from the fishery (Hara, 1996).

In contrast, there is a very high sense of ownership of decisions and regulations on Lake Chiuta due to the participation of vested interests (gear owners) in both actual fishing and management bodies. In the latter case, material incentives have not, so far, been an important factor for involvement in management functions. Another observation is that because of the high proportion of vested interests in both fishing and the BVCs on Lake Chiuta, peer monitoring is much more common on Chiuta than on Malombe. Furthermore, the fishers, led by BVCs, carry out enforcement activities on their own on Chiuta, while on Malombe enforcement of the regulations by fishers is rare.

Two important considerations are derived from the above discussion. Firstly, *who* participates in the management bodies and in the management process is vital to the success of the BVCs. Secondly, what are perceived as incentives for participation are crucial factors for co-management arrangements. The greater the participation of vested interests, as is the case on Lake Chiuta, the greater the possibility of creating effective management bodies. On Lake Malombe, the involvement of crew members, who make operational decisions out on the lake, will be particularly important if BVCs are to function effectively as management institutions.

Negative effects of short term external (donor) funding

The introduction of co-management on Lake Malombe/Upper Shire River was implemented as a multi-donor funded programme (Hara, 2001; Fisheries Department, 1993). This has prompted the usual concerns surrounding donor-funded programmes, such as the dependence syndrome and the sustainability of activities once the project is phased out. Already, the Lake Malombe/Upper Shire River programme has been severely affected as most of the donor projects that supported the programme have been phased out before the management regime could be firmly entrenched (Donda, 2001; Hara, 2001). Unlike the programme on Lake Malombe, the set-up on Lake Chiuta is run by fishers without outside financial assistance. The only government assistance they received was training during the initial stages of the implementation of the programme. Fishers make contributions towards a fund that is used to pay for agreed items and expenses (Njaya *et. al.*, 1999). They call their own meetings and enforcement is carried out using their own fishing vessels. Because it does not rely on outside assistance, the programme on Lake Chiuta would appear to have a much greater chance of being sustained than that on Lake Malombe. In addition, the impression seems to have been created on Lake Malombe that co-management is a donor-funded government project of limited duration, rather than a long-term partnership that should move towards self-sustainability. Even the incentives for participation (in Lake Malombe) are largely seen as being monetary, especially for BVC members.

Ineffective and ambiguous revised legislation

The revised fisheries act, the Fisheries Conservation and Management Act of 1997 (Government of Malawi, 1997a), contained four particular changes essential for the promotion and facilitation of the participatory approach. These were

(a) the introduction of flexibility to allow for a regular review of policy and regulations;
(b) the transfer of property rights over specified fish resources to communities;
(c) permission to allow a part of the revenue obtained from gear licensing to be ploughed back to local-level institutions to cover administrative costs and incentives; and
(d) to provide for the transfer of management responsibility to local institutions when appropriate. One of the objectives of the revised Act is stated as being '...the establishment of participatory fisheries management and empowerment of communities for the conservation and management of fisheries resources' (Fisheries Department, 1997:4) and that it will be the duty of the Director of Fisheries to promote participatory fisheries management. S/he will do this by

'...facilitating the formation of BVCs and also he will facilitate the committees' role in making by-laws for fisheries management in their areas. In addition, he will ensure the empowerment of local communities in the control and management of special fishing areas' (Fisheries Department, 1997:5). The new Act provides for the recognition of associations representing fishers and fish traders and also, the delegation of powers to Honorary Fisheries Officers to enforce regulations (Fisheries Department, 1997). After all these hints of real devolvement of power and responsibility, the Act goes on to say that 'the new Fisheries Management Authorities and BVCs shall be protected and managed in the manner advised by the Fisheries Department for their benefit' (Fisheries Department, 1997: 9). Whether 'empowerment' and 'promotion of participation' mean legal recognition of local institutions created for the purposes of fisheries management is a matter of conjecture. The guide to the act concedes that '...the authority of the fisheries management groups is not specifically mentioned in the Act' (Fisheries Department, 1997: 9). It would appear, therefore, that some of the provisions in the revised Act might be open to subjective interpretation. Thus, while the revised Act seems to provide adequately for most of the needs of a co-management regime, there are still some aspects in which it seems to fall short. Two of these are the *real* transfer of property rights; and the ability of local managerial institutions to prosecute offenders and apply sanctions. The legal transfer of property rights is particularly important for implementing limited entry/access in order to reduce capacity and effort in fisheries, such as Lake Malombe, which are already over-subscribed. The Act does not seem to make adequate provision for such a transfer. Because the initiative for control came from the fishers themselves, the need for legislative backing in order to limit *nkacha* fishers does not seem to be required at the moment on Lake Chiuta. The problem is that village headmen on the Mozambicuan side of the lake are apparently giving permission to *nkacha* fishers to fish from their villages.

Changes to the powers of various courts under the new judicial system have the potential to hamper the ability of traditional courts to prosecute and apply sanctions (Hara, 2001). Under the new system, criminal offences can only be heard at the Magistrate's Court, while Traditional Authorities can only hear civil cases. In this context, it is not clear how Traditional Authorities can adjudicate over fisheries offences, which are criminal, when they are only allowed to preside over civil cases. Furthermore, village headmen cannot impose cash fines under the new system. Such ambiguities could affect the implementation of community-based regimes. These shortfalls suggest that the revised Fisheries Act might still be inadequate in providing all the provisions that local-level institutions might require in order to carry out their assigned tasks efficiently and effectively under a community-based regime. The main problem with the revised Fisheries Act, though, remains that of implementation. Two years after it had been passed, most of its provisions remained unimplemented. This has affected the legal authority of Beach Village Committees (BVCs), the legality of enforcing the revised regulations, and the legality of ploughing back a part of licensing revenues to communities.

The effects of prevailing socio-economic conditions

Lake Malombe lies in Mangochi District, which is characterised by adverse socio-economic indicators, such as high unemployment and population growth, inadequate land size holdings, low agricultural productivity, high rates of illiteracy, low self-help voluntary spirit, and weak micro-enterprises (Government of Malawi/United Nations Development Programme, 1998). In 1996, an estimated 75% of the people in the economically productive age bracket did not have formal, full-time, employment (Government of Malawi/ United Nations Development Programme, 1998). The inability of seasonal agriculture and the formal sector to absorb most of the economically active proportion of the population means that most people obtain their incomes from fisheries. The fact that the number of crew members continues to increase, even though the number of fishing units had been declining during most of the 1990s (Hara and Jul Larsen, forthcoming) is a clear indicator of this trend. This absorptive role that the fishing industry has to play is placing the fisheries under ever increasing pressure.

One of the underlying reasons for the government to introduce co-management on Lake Malombe/Upper Shire River was that this could facilitate the introduction of measures to limit access and entry into the fishery, thereby reducing overall fishing effort. In the context of Lake Malombe and the Upper Shire River, such measures were seen as desirable in principle but unpalatable in the short-term due to the economic hardship and possible social disruption they could bring about (Bell and Donda, 1993). In addition, there are moral and strategic dilemmas for fishers as far as proposals to limit access are concerned. Fishers are not keen on the proposal to introduce limited access because it implies the 'privatisation' of a common pool resource in which everyone has historically been free to fish (Hara, 1996). In addition, such measures imply denying others in similar desperate economic circumstances the chance of deriving a livelihood from the fishery.

Platteau (1992) analyses this anomaly from a perspective of the problem of privatising open fishing space when population pressure makes this imperative. The development of such rights, he argues, does not happen in fisheries because the possible procedures for doing so entail prohibitively high administrative costs and are not usually politically feasible. Jentoft (1993: 24) captures this moral dilemma accurately in what he terms the 'lifeboat dilemma': 'What is to be done when the lifeboat is full? Should one more be taken aboard at the risk of sinking, or should those aboard row hard to get away from all those crying to be saved?' In addition, fishers have expressed the worry that they risk being excluded from other fishing areas to which they seasonally migrate during the closed season as a tit-for-tat action by other fishers from other areas. Clearly, the success of measures aimed at limiting effort is dependent on the ability of the general economy to act as a sink for excess labour from the fishing communities. So long as employment opportunities in the other sectors of the economy remain low and the fishery continues to act as an employer of last resort, such measures are likely to be viewed unfavourably within the fishing communities.

Unlike Lake Malombe, Lake Chiuta is situated in a rural area. Until the 1990s, Chiuta has remained largely under-commercialised and, compared to Lake Malombe, the population density in the area is low. As a result, the area still has adequate farming land for the local community, and fishers still combine fishing with farming. The fishing community is not, therefore, solely dependent on fishing and the fishery does not act as an economic activity of last resort. The prospect of limiting entry has thus not arisen amongst the local fishers. Historically, local fishers have used selective gears such as gill nets, traps and long lines, which seem to have ensured stable catches since the early 1990s. Apparently, migrant fishers are not excluded from the fishery, provided they are using gears similar to those used by local fishers. Local fishers agree that destructive seine nets, such as the *nkacha*, must be kept out of the lake.

The rural setting of most fishing communities, their assumed homogeneity, kinship social structures and traditional authority systems of governance would appear to provide favourable conditions for community-based management systems. In reality though, fishing communities such as those on Lake Malombe and, to a lesser extent, Lake Chiuta, are greatly influenced by market economies. While in former times they might have fished for subsistence purposes only, fishers' objectives are now largely to catch fish for sale. The highly commercialised aspect of the Lake Malombe fishery since the 1980s could be explained on such a basis. As improved roads are built to reach Lake Chiuta and the population increases, local communities in the area might experience similar problems, unless they can learn to limit access into the fishery. This remains one of the challenges for co-management on Lake Chiuta.

In the 1990s, the World Bank/International Monetary Fund Structural Adjustment Programme, combined with greater globalisation, has resulted in the collapse of many local manufacturing industries, and the shrinking of the manufacturing sector (National Economic Council, 1998). Unable to find work in the formal sector, most people of working age are being forced to derive incomes and livelihoods from natural resource-based activities such as farming, fishing, selling firewood etc. With an average population growth rate of over three per cent per annum (Government of Malawi/United Nations Development Programme, 1993), this pressure can only increase as the balance between population growth and natural resources grows more and more skewed. For these reasons, communities are finding it increasingly difficult to apply controls limiting the exploitation of fisheries and natural resources in general. It is doubtful whether co-management alone, without other support or complementary measures, can influence fishers to adopt sustainable patterns of exploitation.

Some thoughts on theory and practice

In the West, there has been increasing public distrust of line agency discretion in the management of natural resources (Lawry, 1994). The last two decades have therefore seen growing demands and advocacy for public involvement in environmental issues. The Brundtland Report (World Commission on the Environment and Development, 1987), which argued that communities should have greater access and control over decisions affecting their resources, enshrined this growing movement. Following the end of the cold war in the early 1990s, donors have demanded political democracy and transparency as essential conditions for development aid. Thus, community participation, which is seen as part of the general drive towards the empowerment of formerly disenfranchised local people, has increasingly been one condition for donor aid. In this sense, donors seem to believe that the subsidiarity principle being commonly applied in the West should also be applied in developing countries as this would result in greater accountability at the local level, resulting in improved resource management and, consequently, positive socio-economic effects on user communities (Lawry, 1994). How serious, then, is the commitment of recipient governments to community participation? Are they just going along with the approach in order to get much needed donor money into their cash-strapped departments and programmes? Given such a scenario, it is possible that communities are being forcibly co-opted into participating so as to meet donor-funding requirements. As authors such as Pomeroy (2001, forthcoming), Hersoug and Rånes (1997) and Jentoft (1989) have pointed out, co-management is a process that takes time to get rooted. Thus, if governments and communities are unable to provide long-term commitment towards co-management after donor-funded projects have ended, co-management programmes are unlikely to be sustainable. Hence, when funding for the Lake Malombe/Upper Shire river programme ended, the government could not continue with co-managerial activities at the same level of intensity as during the project. Some fishers on Lake Malombe were heard to say that 'co-management had come to an end.' While the theoretical basis for the introduction of co-management might be strong, its practical implementation in situations where long-term human and financial commitment cannot be guaranteed requires careful prior consideration.

Another concern is the problem of entrenching the democratic principles on which the co-management regime is largely based, in (rural) communities that have been under national dictatorships for a long time and where traditional leaders still retain enormous autocratic powers under existing undemocratic traditional authority systems. Such historical and cultural contexts have relevance in that in a political environment in which people are not used to arguing or challenging government officials or their village headmen, giving people a formal platform for expressing their views and asserting their rights might not necessarily empower them to do so. It is common on Lake Malombe/Upper Shire River for fishers to complain that their views have not been taken into consideration before decisions are taken, even though they might have attended the public meetings at which the specific issue had been discussed. In this sense, what is left unsaid could be very important and might manifest itself in ways contrary to what might have been thought to have been agreed to by the concerned interests at a public forum. If one of the main reasons for the increasing adoption of co-management is to enforce democracy at local levels, it is still not clear if the real objective for community participation is to enforce participatory democracy or to achieve better resource management.

Whether Western-type, community-level, democratic principles and approaches can function successfully and be of benefit to the management of fisheries in developing countries such as Malawi is an important agenda for research. It is important to note, though, that most decisions in the communities are still taken either by consensus or by the autocratic authority of the traditional leaders and not through some type of democratic process based on winner-takes-all voting.

A third concern relates to reliance on what is usually short-term donor funding for what are, essentially, programmes for long-term institutional reform. If governments and donors are serious about the introduction of institutional reforms for community-based regimes, then long-term human and financial commitments will be necessary. The effects of short-term commitments damaging to that abandon programmes in mid-implementation, can be very the trust between the government and user communities. Since the phasing out of most projects in the Lake Malombe/Upper Shire River Programme and the downscaling of managerial activities, BVCs have expressed the view that the Fisheries Department lacks commitment towards the programme.

Finally, one critical aspect of introducing co-management is that the diagnosis of problems is based on models fostered by 'outsiders', in this particular context, the Fisheries Department and donors. The 'outsiders' influence which problems will be considered and it is common for such models to use templates that assume a direct cause and effect connection when analysing a problem (Turner, 1999). In adopting co-management, it was assumed by the Fisheries Department that by making the user community responsible for the management of 'their' own fishery, then the problem of over-exploitation would be rectified by making them morally responsible for their negative exploitation patterns and illegal fishing activities. The guilt they felt could reduce illegal activities and make them more responsible towards the regulations. But problems concerning fishers, and why they choose particular strategies for exploitation, are usually far more complex than such simple characterisations would imply.

Despite the foregoing, the seemingly positive achievements of the user-based management regime on Lake Chiuta provide a valuable lesson. Where co-operation with government is initiated and driven by the fishers themselves, the regime appears to have much greater potential for achieving positive management outcomes in terms of sustainable exploitation patterns and optimum socio-economic benefits for the stakeholders concerned. The Lake Chiuta case gives hope that, given the right conditions and appropriate government interventions, it is possible to introduce viable community-based regimes in Malawi.

Conclusion

In this paper, we have attempted to analyse and draw lessons from the contrasting outcomes of the two co-management regimes in Malawi. We have argued that both internal and external factors have had a great impact on the outcomes of the two regimes. Amongst the internal factors are the organisation of co-management, struggles for power and authority, the level of involvement of vested interests and the incentives for participation, all of which have been crucial to the workability and stability of the arrangement. The external factors that have impinged on the regime include limited financial support, the inadequacy of revised fisheries legislation and, most of all, prevailing socio-economic conditions.

There is little doubt that the increasing over-exploitation of most capture fisheries in Malawi is dangerously eroding the very basis of fishing communities' livelihoods. Whether or not the real solution lies in the institutional reform of management regimes towards co-management is an important question. In large measure, the solution might lie outside the immediate sector as its problems are also related to the macro-economy of the country. The main problem is the narrow economic base of rural communities, with their extreme dependence on the extraction of natural resources such as fish for their economic needs. Broadening the economic base and widening job opportunities will be indispensable to any solutions within the sector. Thus, while co-management might be one positive reform, it will not be sufficient on its own to solve the increasing crisis in fisheries management.

4

Conflicting Agendas in the Development and Management of Fisheries on Lake Malawi

Edward H. Allison, Peter M. Mvula and Frank Ellis

Present debate on fisheries management in Africa centres on the best way to devolve management responsibility to fishing communities (Harris, 1998; Normann *et al.*, 1998). This debate takes place within the context of a perceived crisis in world fisheries brought about by the failure of state-led fisheries governance (Allison, 2001). The success of this new emphasis on community self-management will, however, depend on many factors external to the fisheries sector, including the policy agendas of the various actors in fisheries and rural development at local, national and international levels. This paper seeks to highlight the importance of analysing this policy context when seeking better ways of managing Africa's inland fisheries.

When considering options for development and management, most analyses focus on the key resource characteristics of fisheries – biological renewability, uncertainty of scientific advice regarding the state of fish stocks and absence of property rights governing access (Bailey and Jentoft, 1990). While no one disagrees that these are important concerns, there is a need to look beyond the fisheries sector at wider development policy and practice to understand the constraints and opportunities for development and management.

Lake Malawi provides an interesting case study of contested policy agendas in development and natural resource management. The lake is both an important fisheries resource to the region and a biodiversity 'hotspot', particularly renowned for the spectacular evolutionary radiation of its cichlid fish fauna, making the lake of global significance to conservation and scientific interests (Lowe-McConnell, 1993). Recent Malawian history reflects major political changes in the 20th century – the decline of European colonialism, post-war 'winds of change' and the more recent global move towards political liberalisation. Many of the major donors and international agencies – World Bank, FAO, ODA/DFID and GTZ - have been involved in the fisheries sector in Malawi since independence, bringing a variety of agendas and policies to bear. These development policies are transmitted into management action by a variety of processes that are mediated by institutional and political context. This chapter explores linkages between policy, management and outcomes in the fisheries sector of Malawi, while attempting to identify dominant policy dialogue and the factors that have shaped it since colonial times.

Lake Malawi fisheries: political and economic context

Lake Malawi, Africa's third largest, is shared by Malawi, Tanzania and Mozambique. Its fisheries are of greatest importance to Malawi, a landlocked country with 20% of its territory covered by water. National catches are estimated at over 50,000 t per year, of which between 40 and 60% originates from Lake Malawi (Government of Malawi, 1999). Lake Malawi catches fluctuate between 20–40,000 t per year (Figure 12)[5], with between 85 and 95% of the catch coming from the 'traditional' or 'artisanal' sector and the remainder coming from 'commercial' and 'semi-commercial' (industrial) fisheries[6].

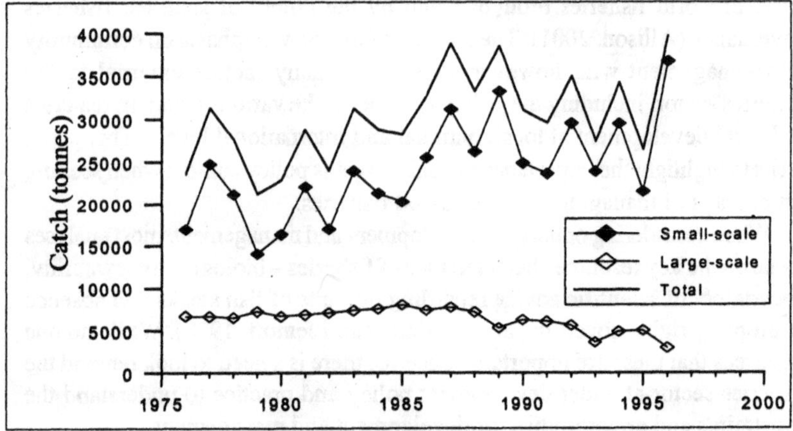

Figure 12: Fisheries production from Lake Malawi, 1976-97 (**Source:** Government of Malawi, 1999). The large-scale fishery is here defined as all pair-trawlers and industrial stern trawlers (the categories 'commercial' and 'semi-commercial' in the Government of Malawi statistics), and the small-scale sector refers to plank boats and dug out canoes (the artisanal sector in Government of Malawi statistics).

5 Lewis and Tweddle (1990) suggest that nocturnal fisheries, particularly for the most important species, '*usipa*' (*Engraulicypris sardella*), may be grossly under-reported in statistics, and that this fishery alone may exceed 50,000 t per year.

6 These terms are problematic – what are known as 'semi-commercial' and 'commercial' fisheries in Malawi are more commonly referred to as 'industrialised' fisheries elsewhere, and the 'traditional' or 'artisanal' fisheries are mostly commercial in orientation. The terms 'traditional', 'artisanal' and 'small-scale' are all commonly in use in the fisheries literature for the small-scale sector and are not regarded as pejorative, although some believe them to be so (Ferguson *et al*, 1993). 'Small-scale' is too context-specific to have acquired a universally accepted definition – in Europe, the largest 'industrial' boats on Lake Malawi would be considered small-scale.

The last ten years have seen an increase of more than 20% in the number of artisanal fishing vessels – dugout canoes and plank boats (Figure 13). More than 53,000 people are involved in the catching or harvesting sector to varying degrees, with an estimated 20,000 employed in processing, marketing and fishing-related businesses such as boat-building and fishing gear-supply (Njaya and Chimatiro, 1999).

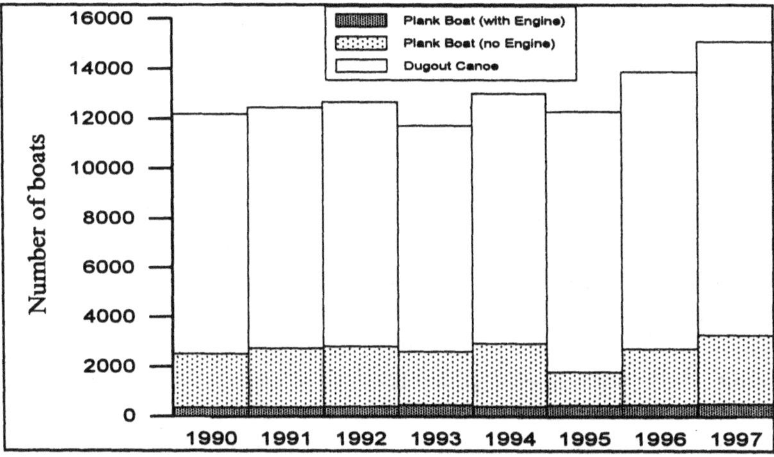

Figure 13: The number of fishing vessels in the artisanal fishing fleet on Lake Malawi, from annual frame surveys.(**Source:** Government of Malawi, 1999).

Men dominate fish catching activities, although women participate in shore-based fishing activities, while the fish processing and marketing sectors involve both women and men. The artisanal fisheries sector is itself tremendously diverse. Boat and gear owners, crew members or 'fishing assistants' exemplify the differing economic groups, the variety of fishing technologies in use is considerable, and diversity of fish species being caught is tremendous (Turner, 1995; Government of Malawi, 1999); there are also ethnic differences between and within lakeshore communities. Social-scientific analysis of the sector has been somewhat lacking (Quan, 1993), but a number of studies seeking to provide policy-relevant analyses of the characteristics of small-scale fishing are currently in progress.

Although there is no evidence of overall decline in catches (indeed catches show an upward trend, with fluctuation, over the last 25 years – Figure 12) there has long been concern that certain inshore fisheries - and those in the south of the lake in particular - are fully or over-exploited (Lowe, 1952; Turner, 1995). This has led to a sustained effort to try to develop fisheries further offshore where fishing effort is presently negligible (Menz, 1995). Production from other lakes has been very erratic, and in the case of Lake Malombe is verified to have crashed due to over-fishing (FAO, 1993). Taken together with population increases, national per capita-supply of fish has therefore dropped from 12-18 kg/yr in the 1970s to around 6-7 kg/yr in the late 1990s. The prevailing narrative is one of increasing population pressure on 'open access' resources and lack of alternative sources of livelihood as the driving forces for the problems in fisheries management (Coulter, 1999). It is these concerns for livelihood and nutritional security that motivate much current fisheries management intervention in the artisanal sector.

The development and management of fisheries has taken place in the context of considerable political and economic change. Malawi was a colony[7] of the United Kingdom from 1891 until independence in 1964. From 1964 to 1994, Dr Hastings Kamuzu Banda ruled in a one-party political system, characterised in later years by authoritarian rule and repressive action against political opponents. The transformation to multi-party democracy, driven by pressure from the international finance agencies, was part of the global tide of political liberalisation that swept through Southern Africa in the early 1990s. In 1994, multi-party elections contested by three main parties were won by the United Democratic Front under the leadership of Dr Bakili Muluzi, who retained the presidency in the 1999 general elections. Despite the extensive political changes, Malawi retains many structural features of the colonial administration. There are 31 District Commissions, with increasing responsibility under a current decentralisation initiative (Government of Malawi, 1997b) and a parallel 'Traditional Authority', headed by hereditary chiefs, whose territories are divided into villages represented by hereditary village heads (men or women). The main function of the traditional leaders since colonial times has been to facilitate all government activities undertaken at local levels, handle land allocation issues and other disputes using customary law, and to oversee all public and traditional ceremonies.

After an early post-independence period of economic growth based on large-scale export-orientated agriculture, difficulties arising from structural and exogenous factors in the late 1970s and early 1980s prompted an increase in donor support and the adoption of structural adjustment policies under the aegis of the World Bank and the IMF. Policies arising from this programme included the liberalisation of trade, reform of the civil service and privatisation

---

[7] First as the Nyasaland Protectorate, then as part of the Federation of Northern Rhodesia and Nyasaland

of state-owned companies or institutions. The result was some success with inflationary control in the 1980s, slowing inflation from 31.4% in 1988 to 11.5% in 1990, but failure to reverse the decline in government revenue (averaging 6.1% annually). The Malawi economy remains fundamentally fragile and vulnerable to factors such as three years of successive drought in the early 1990s and the flotation of the Kwacha in 1994, which led to considerable increases in food and commodity prices. The annual inflation rate rose from 12% in 1991 to more than 50% in 2000 (Government of Malawi, 2000). Except for coffee, Malawi's traditional exports – tobacco, tea, sugar and cotton – have suffered as a result of low output and/or falling prices on world markets (Chilowa, 1998). Tobacco has remained the main foreign exchange earner, but the current depression in world prices paints a bleak future for the country's capacity to earn foreign exchange.

Malawi's main development problem remains absolute poverty, with an estimated 65% of the population living below the poverty line (Government of Malawi, 2000). In response to widespread concerns about the quality and quantity of social sector expenditure under structural adjustment, the Government placed poverty alleviation at the core of its programmes in 1994 (Chilowa, 1998). The extent to which these changing circumstances and policies are reflected in fisheries policy are examined in the following sections.

Fishing for the policy agenda: a history of management and development interventions in Malawi's fisheries

Traditions and tea planters: Colonial policy and its legacy

Evidence of 'traditional' fisheries management in Malawi's pre-colonial period is scanty (reviewed in McCracken, 1987; Chirwa, 1996) but suggests that attempts to limit access to fishery resources in southern Lake Malawi were an instrument used by traditional authorities or rival ethnic groups to exercise power, rather than a measure to conserve fish stocks. In northern Lake Malawi, however, Msosa (1999) reveals a conceptual basis for a resource conservation ethic among the Tonga people, and observes that many traditional fishing gear designs (and the way the gear was deployed) indicate an appreciation of the conservation imperative in fisheries. Survival of 'traditional' conservation measures into colonial times are noted by Lowe (1952) who encountered local rules made by the Native Authorities (e.g. banning the use of seines near river mouths at certain times of year, presumably to conserve spawning runs of fish or nursery grounds for juveniles). The conclusions of most analyses, however, are similar to those made by Lowe (1952: 53): 'Apart from such local rules, any type of gear...may be used at any time and in any place by any African fisherman'. Chirwa (1996: 353) adds 'there are historically, traditionally, and naturally, no tenurial rights on the lake'. The prevailing view running through narratives on Malawi fisheries is that they are open access resources whose ultimate fate, without external intervention or internal reorganisation, would be a 'tragedy of the commons' (Bland and Donda, 1994).

For the first 40 years of colonialism (1891-1931) the government made no attempt to regulate or stimulate development of the fisheries sector in the Protectorate. The colonial view was that:

> 'No interference, commercial or otherwise, could be of any benefit to the native, who makes his own string, his nets and canoes and his basket traps [and who] puts out his nets when the Lake allows him to do so' (Dr. C. Christie, British Museum Nyasa Expedition, 1935, cited in McCracken 1987: 423).

Despite this lack of support, the fisheries sector in Malawi grew through the first decades of the 20th century and it is clear that by the immediate post-war (1945-47) period, small-scale fishermen were already orientated towards commerce, rather than subsistence (Lowe, 1952). In the 1930s, the organisation of this commercial fishing began to change from an essentially African activity, to one that also involved European and Indian entrepreneurs. Early colonial interventions that appeared to support small-scale fisheries over the larger-scale operations of non-Africans were restricted to what was necessary to maintain control without compromising more important sectors of the agricultural economy (McCracken, 1987). Such trade-offs between fishing and more important economic interests are a recurrent feature in the fisheries sector around the world (Hanna, 1999: 284).

The fisheries policy of the colonial administration in the 1940s and 50s was basically reactive to the contradictory agendas of different interest groups within and outside fishing. The first interest group were the tea planters of the Shire Highlands, who needed to provide regular supplies of cheap dried fish to estate workers and their families to retain their labour forces. A persistent policy emphasis on meeting domestic nutritional demand was thus established, and brought the tea estates and the colonial government that supported them into conflict with the non-African fishing interests that were looking to supply more profitable export markets (McCracken, 1987: 425). Another interest group were African 'small-scale commercial' fishermen and fish traders, predominantly returning labour migrants whose sources of economic power came from non-traditional avenues of wealth and who tended to break away from the control of the traditional leaders. These fishers also received support from nationalist politicians, many of whom came from the lakeshore and also had fishing interests (Chirwa, 1996). Thus, in addition to the tea planters, there were three groups within the fishing industry competing for influence:

the non-African commercial fishers; the African entrepreneurs and the 'traditional' subsistence-orientated fisherfolk resident in lakeshore districts.

Traditional fisherfolk were often part-time fishermen, with land-holdings along the lakeshore. Studies of the fisheries sector of this period (Lowe, 1952; McCracken, 1987; Chirwa, 1996) all emphasise the intimate connections between fishing and other sectors of the economy. Linkages between fishing and other sectors of the economy assume particular importance in the context of institutions regulating access to natural resources on the lakeshore. Attempts by the colonial government to assume some control over the lake's resources resulted in the annulment of the historical and natural rights to the lakeshore and further compromised the position of the traditional authorities (Chirwa, 1996). The legacies of this intervention are conflicts over land rights, particularly on fishing beaches on the southern lakeshore, and the disenfranchisement of fishing communities as tourist and weekend-cottage development occupies much of the beach-front land along southern Lake Malawi (Derman and Ferguson, 1995).

Conservation interests also began to be heard from the 1930s onwards. The concerns of colonial officials focused on the design of 'native' nets that did not allow small and immature fish to escape, and the weir traps that were 'a disastrous practice ... catching every fish in the river' (Chirwa, 1996: 362). What colonial administrators had perhaps failed to see was that the timing and location of use of these fishing methods made them more selective and less destructive than might be thought. Chirwa argues that the strongest reason motivating intervention to implement conservation measures for 'traditional' fisheries was not based on scientific evidence of fish stock decline, but on distaste for the methods Africans used to harvest fish, viewing them as 'destructive, primitive, unprofessional and unsporting' (Chirwa, 1996: 363). This contrasts markedly with the respect for the skill and knowledge of traditional fishermen shown in the writings of some of the pioneering biologists of the time (Worthington, 1996; Fryer, 1999).

Recent assessments of the achievements of the colonial government fall somewhat short of unqualified praise. Their policies are described as 'feeble and contradictory' (Chirwa, 1996: 352), their attempts to implement these policies deemed a 'failure' and their ability to intervene to support African commercial fishing 'negligible' (McCracken, 1987: 429). Perhaps not sufficiently recognised by these authors was substantive scientific work (Bertram *et al.*, 1942; Lowe, 1952; Jackson *et al.*, 1963) undertaken during the colonial period. Motivated by concern for the nutritional status of people in Malawi, and by the need to provide income and employment to rural people, these studies laid the foundations for what could have been a more effective fisheries policy. They certainly did much to assess the status of the stocks and prospects for further fishery development.

As Chirwa (1996) concludes, no single group won the contest for the fishing rights and control over the environmental resources of Lake Malawi but the greatest losers were the traditional leaders: chiefs and village headmen whose authority was undermined by the colonial government's attempts to use them to implement its policies. Resistance to the colonial state was therefore extended to resistance to traditional authority. The traditional authorities were also used by the Banda regime to implement government policy at local level, and there too, resistance to the Malawi Congress Party (MCP) involved defiance of traditional authority. A key lesson for current attempts to institute community-based management that emerges from this analysis is that traditional leaders have already been used as the instrument of state control by two non-democratic regimes, and their authority in both cases was undermined. It is thus important to examine carefully the basis for current partnerships between traditional authority, local government and resource users, including the way these relationships are specified in recent fisheries legislation.

Development policies 1964-1994

The Banda regime was characterised by a high degree of central government intervention in all aspects of Malawian life, and by policies that showed a clear preference for industrial-scale development. The Malawi economy grew impressively during the 1970s due to a combination of the expansion of exports from estate-based agriculture, growth in the manufacturing sector through import substitution industries and increased remittance earnings from migrant workers to other Southern African countries (Kydd and Hewitt, 1986). Government policy gave a small class of estate owners preferential access to land, credit and tobacco markets (Kutengule, 2000, for review). In contrast, smallholder production and marketing were negatively affected by the operational biases of the state-owned Agricultural Development and Marketing Corporation (ADMARC), which purchased small farmers' crops at set prices (Harrigan, 1988). The result of these policies was that the share of smallholder produce in total exports declined from 68.7% to 33.8% between 1969 and 1978 (Tellegen, 1997). Population pressure on limited customary land and smallholders' vulnerability to food insecurity and poverty were increased, especially in the less agriculturally productive areas, and among the poorest people (Kydd, 1984; Lele, 1990). Development that favoured the operations of a few private monopolies and the parastatal Press Corporation (in which Banda was a major shareholder) and strict licensing of new enterprises constrained the development of a local entrepreneurial class (Tellegen 1997).

Thus, Malawi adopted a 'dual economy' approach (Lewis, 1954) at a time that rural development policies based on small-scale farming had already been introduced elsewhere. This resulted in a legacy of development policies and interventions that tended to lag behind experiences elsewhere (e.g. Smale,

1995). These retrogressive policies led to a concentration of the labour force in an unproductive smallholder agriculture that received no policy support, and whose opportunities for income generation were severely constrained by official bias in favour of large-scale enterprises.

The rapid economic growth of the 1970s proved not to be sustainable, due to a combination of poor policies and external shocks. Reduced agricultural productivity resulted from droughts, managerial inefficiency in tobacco estates and diminishing profitability of smallholder agriculture attributable to ADMARC's extractive pricing policies. This was compounded by the disruption of trade routes through Mozambique during the civil strife in that country, oil price rises, declining terms of trade for agricultural exports and mounting fiscal and balance of payments deficits. The influx of Mozambican refugees put a further burden on Malawi's resources, as did the cessation of international labour migration after majority rule in South Africa reduced the inflow of remittances and foreign-exchange earnings that had partly sustained the rural economy (Kutengule, 2000). Structural adjustment and development policies in the 1980s were introduced to deal with these problems. Their impact on the performance of the economy is controversial, but some analysts conclude that the gains from the adjustment process were too little to off set the losses among the poor (Kandoole, 1990; Kaluwa *et al.*, 1992).

'Modernising' the fisheries

In seeking to analyse the extent to which wider development policy in Malawi was reflected in fisheries development policies of the time, we find little of help in the rather vague 1973 fisheries policy, the main goals of which were:

(a) To manage the fisheries of natural waters of Malawi so that the optimum sustainable yield is obtained ('optimal yield' is not defined further).
(b) To foster the establishment of a stable and economically viable fishing industry through the development of the activities of the present fisheries.
(c) To assist the efficient landing, processing and distribution of all fish caught and to ensure the achievement of maximum landing is not hampered by lack of marketing facilities.
(d) To encourage fish culture as a means of supplementing supplies of fish from natural waters.
(e) To protect Lake Malawi's and other waters' endemic fish fauna as a scientific and educational asset.

Fisheries Department targets to meet these objectives are stated in terms of maximising sustainable yields, reducing post-harvest losses, increasing economic efficiency of fishing enterprises, promoting fish farming and diversifying fish products. In essence, it was a production-orientated modernisation agenda, concerned with increasing economic efficiency of harvesting, processing and distribution of fish by introducing Western fishing technology, and reorganising the decentralised marketing structure. It reflected thinking in fisheries development at the time (Cycon, 1986), which seems to have lagged behind that in other fields of development (Ellis, 2001). The concern for the lake as a scientific and educational resource is likely to have originated from one of the expatriate fish biologists or freshwater ecologists working in the Fisheries Department.

The conventional development wisdom of the 1950s to 1970s was that the cause of persistent poverty among small-scale fishermen lay in the limited productivity of small boats and fishing gear, a problem curable though provision of more effective technologies (Bailey *et al.*, 1986). There was also a focus on organising fishers into co-operatives to take advantage of economies of scale and to provide a means to deliver large-scale, centralised technologies aimed at increasing the efficiency of artisanal fisheries or developing an export industry. Development efforts in the fisheries sector in Malawi reflected both these trends and were implemented largely through state-led programmes supported by the UK aid programme. Expatriate staff seconded to the Malawi Government at the time included boat designers, marine engineers, fishing gear technologists and fish processing experts.

Post-independence development assistance to the fisheries sector appears to have been problematic. Watson (1987) alludes to defunct ice-plants and pair-trawlers having difficulty maintaining their Norwegian diesel engines. Attempts to create a model for a centralised marketing system proved a failure: a 'fish landing centre' at Lifuwu, on the south-west shore of the lake was located in a position that was too open and exposed to provide a landing for the pair-trawler fleet. Attempts to mechanise the artisanal fleet to extend their range of operations and speed up the delivery of fish to lakeshore markets also seems to have failed:

> 'Escalating fuel costs and difficulties in obtaining spare parts for outboard engines as well as inboard diesels have seriously hindered the vessel mechanisation programme. Many of the plank-boats have reverted to using oars or paddles' (Watson, 1987).

While catch monitoring data indicated the size of the fishery was increasing, the amount of revenue collected from fishing licences was actually decreasing as a shortage of operating funds meant that Fisheries Department officials were unable to check on fee payments. There was also 'a serious degree of disregard of the regulations by both commercial and artisanal fishermen' (Watson, 1987: 10). The extension service was under-funded and generally ineffective as it received insufficient guidance on extension messages. The Fisheries Department was also burdened with the running costs of donor-funded developments such as ice plants, which usually ceased to function after donors withdrew.

Paradoxically and despite this analysis, Watson (1987) recommended additional ice plant, chill storage and beach-landing developments, feeder roads and improved market facilities for wholesale traders. Some of these suggestions were later taken up by the World Bank Integrated Fisheries Development Project (Ferguson *et al.*, 1993). The calls for more infrastructure developments at strategic fish landing sites and main centres contrast with Cycon's (1986) contemporary plea for more appropriate development that took account of the decentralised nature of small-scale fishing operations and their associated social structures.

The technology-transfer approach produces well-known problems. Profit and production-oriented fisheries development strategies promoted by international donors and national policymakers have rarely considered local community aspirations and the impact of technological innovation on community social structure (Bailey *et al*, 1986). In heavily exploited fisheries, new technology benefits some at the expense of others, and thus often causes disequalising distributional impacts. This is not to suggest that new technology itself is bad, but the socio-economic context for its introduction must be well understood, a factor acknowledged at the time in agricultural development, as manifested by the development of farming systems research approaches[8]. There was no system in place to assess the impact of technology transfer on Malawian fisheries and not enough knowledge of social and economic structures to interpret their likely impact. Experience elsewhere suggests that fishery projects designed to make fish processing and distribution more efficient by replacing private fish buyers with formally organised, male-dominated co-operatives have tended to undercut the primary source of independence for women in fishing communities (Bailey and Jentoft, 1990: 341). It is perhaps fortunate, therefore, that the attempts to organise fish marketing in Malawi failed.

It can also be noted that small-scale fishermen themselves are constantly engaged in technological adaptation and innovation. An example in Malawi is the development of the open-water *chirimila* seine, developed by the Tonga fishers from the north and spread to the south by migratory fishermen (Hartmann, 1987).

Some externally-driven technology transfer efforts were successful, however. Plank boats of around 4-5 m length (flat-bottomed so that they could be hauled up on beaches) were built and promoted by the Fisheries Department from the 1960s to substitute for dugout canoes. They provide a larger, more

---

[8] Farming systems research approaches have been used in Malawi for aquaculture development, where the policy to support integrated agriculture-aquaculture for smallholder farmers, using indigenous species, reflects the 'agriculture first' and environmental agendas of ICLARM, the principal agency behind aquaculture development in Malawi since 1987.

stable platform for the exploitation of a variety of the lake's fisheries, while being sufficiently small to be powered either by paddling or small outboard engine. The introduction of nylon gill-nets improved the efficiency of this fishing method, but perhaps with subsequent resource conservation implications (Njaya and Chimatiro, 1999). There have also been successes in the post-harvest sector, with projects to introduce more fuel-efficient and lower temperature smoking kilns from West Africa (Njaya and Chimatiro, 1999) and the promotion of 'actellic' powder to use on sun-drying fish to reduce insect infestations (Hartmann, 1987). Development assistance also supported the construction and operation of fisheries and aquaculture research centres and a fisheries training school within the Fisheries Department.

In reviewing and proposing amendments to the fisheries component of the 1986-1995 National Development Strategy, Watson (1987: 32-33) endorses a policy that combines technology transfer, centralisation, and support for large-scale development. There are no social goals for fisheries in the development strategy, although a socio-economics consultant on Watson's team proposed that these be specified in the policy. These goals were: increased employment and income, improved marketing, more equitable distribution of fish supply inter-regionally, and provision of fish supplies to low income populations (Hartmann, 1987).

President Banda's commitment to centralised, state-led modernisation policies reflected orthodox development thinking in the 1950s. This model for development was already being called into question elsewhere in the 1960s (see Ellis, 2001, for review), but persisted until the multi-party elections in 1994. Centralised, production-orientated modernisation policies also persisted in the fisheries sector until the last decade (see Le Sann, 1998, for review). Thus, both national development strategy and international fishery sector interests coincided, in Malawi, to favour the type of modernisation policy described above, well after such approaches had been abandoned elsewhere.

Fishing in deep water

A persistent theme in fisheries development for Malawi has been the attempt to move fisheries further offshore, to alleviate pressure on inshore stocks (important to both fishing and conservation interests) and to target under- or un-exploited resources. Surveys of the offshore pelagic zone in 1953-55 (Jackson *et al.*, 1963), 1977-82 (FAO, 1982) and 1990-94 (Menz, 1995) have all failed to provide impetus to offshore fisheries development. This is because the fish that are found offshore, although theoretically able to sustain a fishery of up to 34,000 t per year (Thompson and Allison, 1997) are mostly too scattered, deep-dwelling and of low value to make fishing economically viable at present, by either artisanal or industrial technologies (Menz and Thompson, 1995). There have also been several attempts since the late 1960s to seek out new bottom-trawling grounds to enable the expansion of the industrial fishery (e.g. Tarbit, 1972; Thomasson and Banda, 1996; Kanyerere, 1999), at present confined largely to the south-east arm of the lake (Turner, 1995). With the exception of the 1990-94 survey, which specifically assessed the potential for artisanal fisheries expansion as well as industrial (Menz and Thompson, 1995) these studies and proposals can all be interpreted as supporting expansion of the large-scale or industrial fishing sector. In most industrialised fisheries elsewhere, this type of exploratory fishing is nowadays financed by the private sector.

The government and donor support for offshore fisheries expansion aligns broadly with the themes of production-orientated modernisation and supply of fish to consumers. Once again, the interests of small-scale fishing are peripheral to these other agendas.

State conservation policy and fisheries

The potential expansion of trawling, the mooted introduction of exotic fish to boost the low offshore pelagic fish production (FAO, 1982) and the increasing fishing effort in the artisanal fishery all galvanised conservation concerns. The most tangible outcome of this concern was the designation, in 1980 (with WWF support), of an aquatic National Park centred on Cape Maclear in the southern part of the lake. The park was designed to conserve the rocky littoral habitat (up to 100 m from shore) of the '*mbuna*' cichlids, not targeted by fisheries and found in the shallow littoral zone of the lake. The park protects the adjacent forested catchment, and certain offshore islands, but did not displace 'enclave' fishing villages occupying the sandy beaches within the park area. The park appears to be fairly successful in reconciling conservation aims with development, and attracts considerable tourism, some of the benefits from which accrue to enclave villages (Coulter, 1999). The park offers no protection, however, to sand-dwelling species and offshore fish vulnerable to fishing activity (Turner, 1995), yet it is the conservation of these fish that are imperative to the future of small-scale fishing. Conservation activities thus reflected the priorities of the international community, rather than those of resource users, although their interests were, to some extent, protected.

Also significant to conservation and development was the rejection of the notion that fish yields could be boosted by the introduction of clupeid (sardine-like) fish from Lake Tanganyika. The move to reject it was led by conservation interests that had witnessed the loss of cichlid species in Lake Victoria following the Nile Perch introduction and feared for the fate of the endemic cichlids and the fisheries they supported (Barel *et al.*, 1985). It was reinforced by subsequent studies of the lake's ecology that found that production was unlikely to be enhanced by such an introduction (Allison *et al* 1995, 1996). Species introductions were a popular 'technical fix' to increase production in inland fisheries during this period (Petr and Kapetsky, 1983), but had serious consequences for endemic biodiversity and for the socio-economic structure of fishing communities (see the papers in Pitcher and Hart (Eds) 1995: 1-238 for review of the Lake Victoria experience). Malawi has been 'a shining example to other countries in its recent avoidance of introductions of exotic species and in its development of legal instruments to protect its natural environment (particularly Lake Malawi)' (Pullin, 1991: 1). Once again, the key actors in the debate over species introductions in Lake Malawi were international fisheries scientists and conservationists (see Allison *et al*, 1996, for references).

The Life-President's fish

The donor and departmental resources devoted to exploratory fishing on behalf of industrial fishing interests is just one instance of the outcome of policy support for the industrial sector. With attempts to centralise fisheries operations by provision of infrastructure having largely failed by 1987, most donor-funded landing stations were handling only the fish caught by the Department's own 'research' vessels. These vessels were fishing commercially to pay operating costs, so that their research and development function was subordinated to the need to generate revenue (Watson 1987: 14-15). The sale of fish caught by the government placed the Fisheries Department in a position of competing for markets with the very fishermen that it was intended to serve. This problem was compounded by the manner in which fish selling prices had to be authorised beforehand, meaning that they could not reflect day-to-day changes in market values and inevitably undercut local prices to the detriment of fishermen landing in the same area. Donors have not learnt from Watson's cautionary tale, and the ICEIDA-donated Research Vessel 'Ndunduma', one of the two most powerful fishing vessels on the lake[9] has, since 1993, continued this practice of operating mostly as a commercial fishing vessel to pay its running costs.

The above case raises the important question of who the Fisheries Department is intended to serve – those involved in the fisheries sector or the interests of the wider public as consumers? In Europe and the US, there is a prevailing notion, supported in law, that fish belong to all citizens of a state (Allison, 2001). Fisherfolk are granted the privilege of making a living by being allowed to exploit fish belonging to the state and all its citizens. The Malawi Fisheries Department represents the interests of the state and its taxpayers by ensuring the value of the resource is not diminished by overexploitation, and to ensure that revenues are collected from those allowed

---

9 The R/V Ndunduma's sister ship the F/V Kandwindwe is operated by the Malawi Development Corporation (Maldeco) Ltd.

access to this public resource. It has no primary responsibility for the welfare of fishing communities, which may explain why Fisheries Departments are perceived to put the interests of fish and the wider public before those of fishermen:

> 'The paramount responsibility of the Department of Fisheries will remain the protection of the existing fish resources by means of appropriate research, vigorous application of the various regulations, and the collection and analysis of relevant data' (Government of Malawi. Cited in FAO, 1995c: 1).

Under the 'fish as state-property' view, community-based management essentially becomes a cost-effective means of ensuring compliance with state law. Early proposals to introduce community-based management in Malawi reveal this perception (FAO, 1995c: 33-34), with a vision of such management as a programme of awareness-building and training of 'community leaders' by Fisheries Department extension staff, with the expectation that the leaders would then monitor and enforce compliance with government regulations. Once again, 'traditional' authority was being co-opted to do the state's work. This model does not fall within the currently understood definition of community-based management, where the community has a role in deciding the rules, rather than having them imposed externally and simply being the vehicle for implementing them. Perhaps this was as far as it was possible to go with devolving power under the Banda regime, but it does seem that the FAO lacked confidence in the community-based approach. They immediately followed the recommendation for pilot community schemes to be introduced with a call to strengthen the enforcement and licensing functions of the Fisheries Department (FAO, 1995c: 34-35). This half-hearted and top-down view of community-based management perhaps explains some of the subsequent problems in effective implementation (Hara, 1998).

With fish viewed as state property[10], it is also less important how fish are caught, so that the objective of fisheries development becomes one of maximising the supply of fish to the consumers, as economically as possible. The best way to achieve this, it was thought throughout the Banda period, was through support for large-scale, mechanised exploitation and centralisation of the artisanal fish marketing system. Centralised fisheries are also easier to monitor, manage and control than decentralised ones, reducing the costs of management and providing further rationale for this approach to management and development.

---

[10] That fish are the property of the state was taken literally by President Banda. As the major shareholder in Press Holdings, he had a personal stake in one of its subsidiaries, the major 'commercial' fishing company operating on the lake – Maldeco.

Fisheries development thinking from the 1960s to 1980s never had the same theoretical focus as the emphasis on the efficiency of small-scale farmers that occurred in agricultural policy (Ellis, 2001). Rather, fisheries development focused on modernisation through provision of centralised facilities, and tried to organise individual small-scale producers into larger collectives, so that they could compete favourably with the 'modern' sector, and eventually acquire the means to join that sector or leave the fisheries. Thus, from the 1950s to the mid-1980s, the dual-economy model persisted in fisheries. The prevailing wisdom that small-scale producers are less efficient than large-scale ones is, however, challenged by the continued domination of the fisheries sector by small-scale fisherfolk, despite support for the development of large-scale fisheries. Further support for the efficiency of small-scale fisheries comes from Lake Tanganyika, where in the absence of development intervention, the artisanal fleet out-competed the industrial purse-seine fleet exploiting the same fish stocks (Petit and Kiyuku, 1995).

Political liberalisation and changing policy discourse

Since the early 1990s, there has been a change in the discourse around fisheries development and management in Malawi (documented by Ferguson *et al*, 1993) that would appear to reflect the transition from 'modern' to 'post-modern' thinking in fisheries (Harris, 1998) or 'top-down' to 'bottom-up' approaches in rural development (Chambers, 1983). This transition was enabled by the change to multi-party democracy, itself forced by pressure on Banda to hold multi-party elections as part of an international agenda linking the provision of aid to 'good governance'.

The paradigm shift is not at all clear cut, however, and remnants of old policies and development practices remain, reflected in donor projects and Fisheries Department activities over the last decade. The conflicting policy agendas within the major World Bank 'Integrated Fisheries Development Project' (1991-99) have already been highlighted by Ferguson *et al* (1993), who point to continued support for commercial fisheries rather than artisanal ones, infrastructure development, and support to expand the monitoring and enforcement capacity of the Fisheries Department.

The FAO 'Chambo Research Project' (1988-92) was the first major multidisciplinary fisheries research project on Lake Malawi, marking a move towards a wider fisheries-sector research approach and away from studies confined to stock assessment and fish biology. It provided the first widely available data on fishermen's incomes, distinguishing well-off owner-entrepreneurs with fishing and other business interests from crewmen whose cash income from fishing was less than US$ 7.50 per month (FAO, 1995c). The project also documented the crash of the Lake Malombe *chambo* fishery

(FAO, 1993) and made a number of management recommendations that marked the replacement of the historical right to free access by any Malawian with a system of government-administered licences, and suggestions for piloting involvement of fishing communities in the enforcement of regulations.

This shift in project emphasis was complemented at national level by donor support to improve management planning and policy-making capacity in the Department of Fisheries (e.g. Bland and Donda, 1994; FAO, 1994a). Also significant were long-term GTZ-funded programmes. Initially, these programmes were concerned with development of small-scale aquaculture and improvements in fish processing around Lakes Chilwa and Chiuta (the Malawi German Aquaculture and Fisheries Development project), but this evolved into a programme supporting local-level institutions in the implementation of community-based management (Bell and Donda, 1993). Out of this programme grew the on-going GTZ-funded National Aquatic Resource Management Project (NARMAP) which, as well as continuing to support the development of community-based management, has also sought to influence fisheries policy at national level by working within the Department of Fisheries to promote revision of legislation and policy and to develop capacity in appropriate monitoring, research and extension activities (NARMAP, 1999)[11].

The different donor priorities are reflected in the fisheries programmes they funded during the early 1990s. FAO and ODA/DFID programmes demonstrate a shift from exploratory fishing and biological research, towards implementation of management based on sector-wide studies. World Bank support (1991-99) was in the area of reform of public institutions, structural adjustment and economic liberalisation, but this was allied to a continued 'fisheries development' agenda that included collaboration with ICEIDA to provide two powerful new stern-trawlers for the Fisheries Department and Maldeco, and to conduct exploratory trawling (Thomasson and Banda, 1996). In later years, this programme co-existed with a World Bank GEF project (1996-2000) aimed at providing a scientific basis for lake-wide environmental conservation.

By the mid- to late-1990s, it was clear that there was no clear policy direction behind interventions in the fisheries sector – the Department was involved in donor-financed projects to expand industrial fisheries and strengthen its own monitoring and enforcement capacity, while at the same time the expanded Fisheries Department was devolving management responsibility to local communities through another donor programme. Fisheries expansion, resource management, and biodiversity conservation

---

[11] Revision of Malawi fisheries policy has also been funded by USAID.

programmes were all underway, with aims that were not readily reconcilable with national poverty alleviation policies. The impression gained is that the Fisheries Department, chronically short of funds, was willing to take on any project donors suggested, whether it fitted with national priorities or not, and that donor activities were poorly co-ordinated and uninformed by either a sectoral or national policy.

The Fisheries Department realised that there existed a policy vacuum in the fisheries sector, with the Fisheries Act, formulated in 1949 and re-enacted in 1973, being criticised by the Director of Fisheries as 'outdated, brief and narrow in scope' (Mapila *et al.*, 1998) and lacking objectives for fisheries management that could be read in law. In response, a new Fisheries Act was drafted and implemented in 1997.

Prominent in the 1997 Fisheries Conservation and Management Act (Government of Malawi, 1997c) is the provision for local community participation in the conservation and management of fisheries in Malawi. There is legal provision for the Director of Fisheries to enter into a management agreement with a (community-based) fisheries management authority providing for a management plan and assistance by the Fisheries Department [Part III 7 and 8 – a mandate for co-management arrangements]. At first sight this represents a significant policy shift by the Fisheries Department. Closer examination, however, reveals detailed specification of the powers of search, seizure, demolition and arrest by a government fisheries protection officer and detailed legislation relating to prohibitions and offences, court proceedings and penalties (Parts X – XII). This suggests either a residual attachment to command and control policies, or the influence of a range of donor agencies with differing policy agendas.

The new Fisheries Act claims to empower communities by giving them responsibility for the 'conservation, increase and management of fisheries resources'. It does this by facilitating the formation of Beach Village Committees (BVCs) and allowing members of BVCs to be honorary fishery protection officers with a legal mandate to enforce the Act. The Director appoints these officers and can determine their duties. (Mapila *et al* 1998: 6). Fisheries Management Authorities (FMAs) can be any organisation of local fishermen, commercial fishermen, traders, processors and NGOs who will choose their own leaders. But the new FMAs and BVCs shall be managed 'in the manner advised by the Fisheries Department for their benefit' (Mapila *et al.*, 1998: 9). This relationship is worked out through individual fisheries management agreements between the Director of Fisheries and the management authority. The Minster can make rules to determine who will be given a licence and who will not, but communities can make rules that may become by-laws and be legally enforceable. In short, the Act appears to grant considerable rights to self-determination by fishing communities, while retaining all the powers of the Department of Fisheries. As Bailey and Jentoft (1990: 342) have said, 'In practice, as distinct from rhetoric, governments have been more anxious to establish local organisations that they can control rather than to create viable co-operatives that have local autonomy and independence.'

The Fisheries Act is to be implemented through the Fisheries Conservation and Management Regulations, which update technical measures for fish conservation, relating to gear dimensions, closed areas and seasons and minimum legal landing sizes for different species of fish. These are complex regulations for a complex fishery, but implementing them at national level is likely to be impossible (Turner, 1995). The Regulations also introduce three types of licence – commercial, sport and subsistence. Commercial licences are divided into small-scale (20 hp outboard or less, or un-mechanised), large-scale and live fish (for the aquarium trade), finally recognising that small-scale fishers are also 'commercial'. It is not clear, however, how boundaries between small-scale commercial and subsistence fishing will be set, as most small-scale fishers keep a proportion of their catch, or exchange some catch for other goods and services, rather than trading all the fish for cash.

Current policy – evading hard choices in fisheries development

The 1997 Fisheries Act was published before the formulation of a National Fisheries and Aquaculture Policy, so that policy follows from law, rather than the other way round. It is difficult to discern a strong direction from the rhetoric of the new fisheries policy. The stated objectives of the new fisheries policy are 'to improve the efficiency of all aspects of the national fisheries industry, production and supply of existing fisheries productions, as well as the development of new products to satisfy local demand and potential export markets.' Alleviation of poverty – the key objective of Malawi's National Development strategy since 1994 – is phrased as a 'constraint' to achieving these objectives (p. vi and p3), rather than an objective in itself, but is also presented as an *outcome* of adopting the new policy (p3). A range of sub-sectoral policies relating to extension, research, participatory fisheries management, fish farming, training, enforcement, riverine and floodplains fisheries and fish marketing are also given. These individual objectives and strategies provide guidance on a huge range of specific strategies. While it must be recognised that different management targets can be set for different fisheries sub-sectors, or even different areas of the lake, some overall strategy is important (Hanna, 1999). At present, almost any fisheries development activity could identify an objective and strategy to which it is relevant, thereby potentially propagating conflicting development objectives in the fisheries sector as a whole.

What is missing from the policy is a series of higher-level goals for the sector. The overall objective, as stated, is too broad and vague, and the individual objectives and strategies for the sub-sectors are too specific. Policy adds everything but takes away nothing – it evades the 'hard choices in fisheries development' (Bailey and Jentoft, 1990) by not, for example, choosing between support for decentralised small-scale processing and development of centralised facilities producing products to a standard acceptable to international markets. It is also a policy that supports participatory management but continues to stress the role of government in setting management objectives and enforcing compliance.

Despite being detailed, there are some surprising omissions from the policy. There is no objective relating to the management of the larger-scale commercial fisheries, so that there is no means to decide on the future allocation of support between industrial and artisanal fisheries sectors. The same lack of clarity is evident on objectives for processing, marketing and private sector investment strategy – there is a policy in support of joint ventures with foreign countries, but who are the Malawian partners – small-scale fisherfolk or large-scale ones?

The 1999 Fisheries Policy represents a laudable attempt to incorporate recent thinking in development, but by also retaining all past policies and options and omitting to prioritise objectives, it fails in arguably its most important function – to provide direction. There is no clear sense whether the government views the fisheries sector primarily as a 'safety net' for the landless poor, a nutritional source for the nation, an important component of the rural economy or a source of government revenue. This lack of policy direction seems to extend into higher-level policy. Kutengule (2000: 248) points out that while rural development policies aim to increase agricultural production as a means of reducing rural poverty, they have no coherent approach to non-farm activities.

Fisheries development involves hard choices between incompatible objectives. The Malawi fisheries policy includes the goals of increasing the supply of fish to domestic markets and promoting the export of fisheries products; increasing fisherfolk's incomes and providing new employment opportunities in fisheries. These are all individually valid goals, but are mutually incompatible, as can be illustrated by considering the example of policy objectives to increase production, increase domestic consumption of fish by the poor and increase fisherfolk's incomes (Bailey and Jentoft, 1990: 339). Fisherfolk's incomes can be raised by increasing the catch, reducing production costs, or increasing prices. Increasing catch is not an option in fully exploited fisheries and will in any case increase the costs of production per unit of harvest as the catch-per-unit-effort (CPUE) declines. Improving economic efficiency by reducing costs of production is best achieved by eliminating excess fishing capacity and labour[12], relieving pressure on the resource and making it less costly for those who remain to harvest a given quantity of fish (higher CPUE). If restrictions on access to fishery resources are promoted, however, this will have a negative impact on employment. Price controls in favour of fishermen (e.g. minimum sales prices) are not good for consumers, and would attract additional investment of capital and labour into the fishing industry, leading to a cycle of reduced CPUE and rising costs of production, while price controls to benefit consumers would only do so at the expense of fisherfolk's incomes, and are difficult to enforce, leading to the development of black markets.

The point made in the above example is that hard choices have to be made, to design policies that will support clearly identified priorities and beneficiaries. This analysis of trade-offs between conflicting policy options and priorities does not seem to have informed the current policy, despite the fact that Quan (1993) clearly outlined policy choices for Malawi Fisheries, and the NARMAP project recognises that resource sustainability will require limitation of fishing effort in its project logical framework (NARMAP, 1999). In considering management and development programmes derived from the new policy, it would be desirable to seek the viewpoint of those likely to be impacted, and ensure full consultation with fishing interests over priority objectives. This will require a change in the way policy is made in Malawi. The current drive towards decentralisation (Government of Malawi, 1997b) could give people on the lakeshore greater say in policy formulation than they have at present.

Synthesis

Fisheries policy in Malawi does not currently provide a clear direction for development. We have argued that such direction can be provided by careful analysis of trade-offs between different policy goals that are agreed in consultation with identified stakeholders. It is also suggested that an awareness of broader development policies and the theories that inform them are a pre-requisite to making effective policies that are compatible with national development strategies.

Coulter (1999) argues that the fisheries development of the African Great Lakes has not been driven by policy, so that the current fashion for policy support in favour of practical management intervention is misplaced. We would agree insofar as 'policy' is taken to mean the statements written in the offices

---

[12] Reducing 'overcapacity' in fisheries is a major policy objective of the FAO Code of Conduct for Responsible Fisheries (see Allison, 2001), to which Malawi is a signatory.

of governments with little power to turn objectives into interventions. However, we hope we have shown in this chapter that 'policy' is made and implemented through a variety of channels, including the agendas of major international organisations with considerable influence in donor-dependent economies such as Malawi. Policy, in this broader sense, has certainly impacted the development of fisheries in the region, if only by being inappropriate and unsuccessful. The role that President Banda's policies have played in the agricultural economy of Malawi has been extensively analysed (e.g. Harrigan, 1988) but such analysis is lacking for the fisheries sector. This lack of critical analysis of fisheries policies may have contributed to the perpetuation of inappropriate, ineffective and conflicting development interventions in the fisheries sector.

We have identified four major sets of actors in making fisheries policy – small-scale fisherfolk and traders, national policy makers, international donor agencies and international conservation interests. The participation of fisherfolk in policy formulation has been extremely limited. Evolving ideas come from the top down and have shaped policy while ignoring or subordinating the values of local fishing communities (Bailey *et al.*, 1986). This policy orientation has remained essentially unchanged from the colonial period until the last few years. The new 'participatory' agenda is not exempted – it emanates largely from the new directives of donor organisations.

Biodiversity conservation agendas have been very influential in guiding the development of Lake Malawi fisheries. Policies have incorporated an explicit commitment to conservation objectives since the late colonial period. Existing conservation actions have prevented exotic species from being introduced and have influenced technical measures for fishery resource management. Some of the latter could have had positive impacts on conserving larger, more vulnerable fish species, but have not been effectively implemented (Turner, 1995). Despite this conservation ethic, the recent 'integrated conservation and development' programmes that attempt to link conservation and livelihoods through an analysis of the economic value of biodiversity have had little influence on the thinking and management strategies for biodiversity conservation on Lake Malawi, or indeed the other African Great Lakes (Allison *et al.*, 2000).

It is difficult to assess to what extent the international conservation agenda is 'owned' within the Fisheries Department at present. Analysis of lessons learnt from biodiversity conservation projects in Africa (Hart *et al*, 1998) suggests that the commitment of many African national governments to biodiversity conservation programmes is weak. Such programmes are seen as the external imposition of an international environmental agenda and governments can even be hostile to programmes promoted and managed by external agents that are perceived to favour 'animals and trees over people'. Hart *et al* (1998) conclude that biodiversity conservation programmes are unlikely to be sustainable unless they are integrated into country development strategies, or financed indefinitely by the international community.

Policy makers still poorly understand the nature of the fisheries sector in Malawi. A World Bank document proposing an Environmental Management Project on Lake Malawi (World Bank, 2000) proposes, among other objectives, to 'support a gradual transformation of the current traditional subsistence fisheries sector into a sustainable, locally based, small-scale commercial fisheries sector'. Such a transformation already took place somewhere between 50 and 100 years ago (Lowe, 1952; McCracken, 1987). The ability to prioritise development objectives for the fisheries sector remains constrained by a lack of information on what is happening on the lakeshore. Clearly, there is a need to find ways to ensure that information flows between fishing communities and the Fisheries Department are improved. At present, the information flows are conceptualised in one direction only: from the Fisheries Department to the communities, in the form of extension messages and directives on management.

A second area where understanding of the nature of small-scale fisheries may be insufficiently integrated into policy formulation is in the relationship between fishing and the wider rural economy. Fishing has never been a full-time activity for a large proportion of the people living on the shores of Lake Malawi. McCracken (1987) details how the relative importance of fishing and fish trading in the colonial rural economy were influenced by lake-level fluctuations that inundated farmland and access to new markets or income-generating opportunities. Adaptability and diversity are still prominent features of lakeshore peoples' livelihoods (Allison *et al.*, 2001a).

Studies of the livelihoods of lakeshore people suggests that institutions (formal and informal) that govern access to fisheries resources need to be understood in the wider context of access to natural resources in general (Kutengule, 2000). Instead, attempts to regulate fisheries have invariably taken place in isolation from other resource management interventions, and have not been informed by knowledge of these access regimes. This has led to a lack of harmonisation of policies and programmes in natural resource management, exemplified by the proliferation of externally-conceived local-level institutions with overlapping functions. For example, the functions of village committees for beaches (BVCs), overlap with both natural resource management, and development (Njaya and Chimatiro, 1999). Studies elsewhere in Africa are beginning to address this weakness in current fishery management systems (e.g. Geheb, 1999; Sarch, 2000). A micro-level understanding of livelihood strategies seems essential for the design of appropriate formal institutions to regulate access, whether state-led, community-led or in partnership.

The history of weak and contradictory policies, ineffective and inappropriate development intervention, lack of clear direction in current fisheries policy, and the continued lack of self-determination by fishing people in Malawi all make for rather depressing reading. There is much that is positive, however. Recent donor support for institutional development in the Fisheries Department – for all its shortcomings (Ferguson and Derman, 2000) – is giving rise to a cadre of well-trained and motivated policy advisors within the institution. Political liberalisation in Malawi ensures that a constructive critique of government policy by Malawian academics and policy analysts can now take place. Decentralisation offers new opportunities for local-level governance of fisheries. The current move towards co-management, while not fully committed, is a long way from the neglect of small-scale fishing by colonial and post-independence governments. The agencies involved (principally GTZ at present) are responsive and flexible to the evolving situation.

Fisheries management in Malawi is thus already moving, albeit cautiously, towards new solutions. Malawi's new policy makers remain dependent on donor assistance, but can now have more influence in shaping how that assistance is deployed. The 'voices of the poor' are yet to feature significantly in the policy process, but the evolving co-management regime and move towards decentralised government may facilitate this communication. What remains to be achieved at central government level is a more clearly defined fisheries policy, with a clear sense of who the beneficiaries are and an associated sector-wide investment plan to ensure donor funds are used efficiently, and to prevent donors from driving, or undermining, government policy. This should help reconcile Malawi's obligations to feed its people, provide livelihoods for its fisherfolk and sustain the lake's environment and fishery resources for future generations.

Acknowledgements

This paper is a contribution to two on-going research projects in Malawi: 'Sustainable livelihoods from fluctuating fisheries' (Project R7336), funded by the Department for International Development's Fisheries Management Science Programme, and 'Livelihood Diversification Directions Explored by Research (LADDER)' funded by the DFID Policy Research Programme (R7554) Both these projects seek to complement macro-scale analysis with micro-studies on rural livelihoods, including those of fisherfolk, to inform the design of policies that enable people to sustain their livelihoods. EHA's work on the fisheries of Malawi during the last 10 years has also been supported through an ODA Associate Professional Officer award, and a grant from the INCO/DC programme of the European Union (Lake Malawi Demersal Fish Project Contract ERBIC18CT970195). We thank M. Kutengule for his useful input and L. Mathieu and J. Mims for bibliographic assistance. The opinions presented in this paper are our own, and do not necessarily represent those of DFID or other organisations and individuals acknowledged here.

5

Contesting Inequalities in Access Rights to Lake Kariba's *Kapenta* Fisheries: An Analysis of the Politics of Natural Resource Management

Kefasi Nyikahadzoi,

The pelagic fishery of Lake Kariba is shared between Zimbabwe and Zambia (Figure 14), and exploits a freshwater sardine (*Limnothrissa miodon*), the '*kapenta*'. In shared resources, how the total allowable number of fishing units is determined is not just a question of following the scientific advice of national or regional scientific institutions, but is also based on the demands of

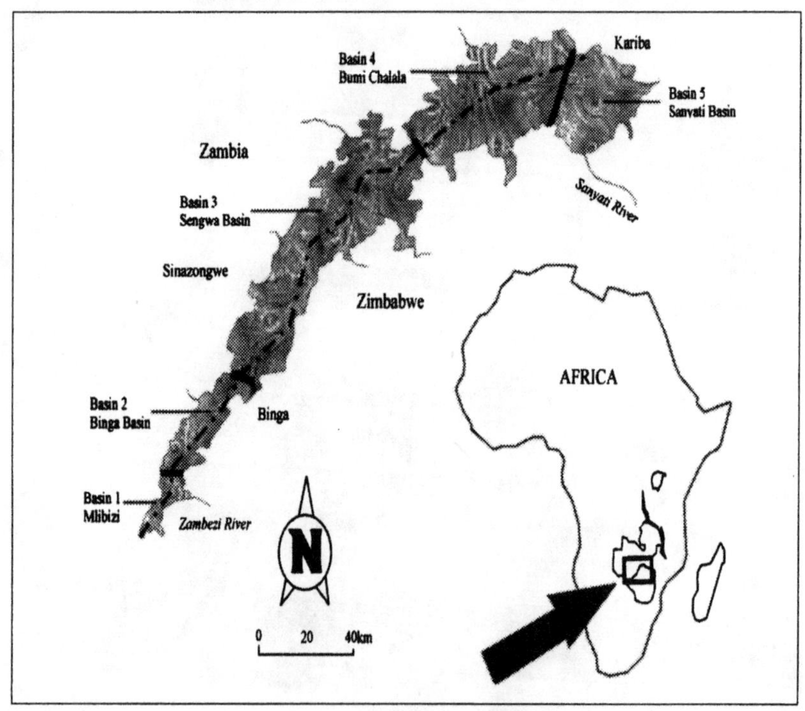

Figure 14: Map of Lake Kariba, Zambia and Zimbabwe

the other countries sharing the resource. The total allowable number is most often arrived at through negotiation between the lacustrine states. Once the number of units per country is determined, each country takes into account its historical, political, industrial structures and social conditions to decide on the allocation of access rights to its nationals. The decision on the number of fishing units that a fisher can own involves different interest groups, each fighting to obtain its share of the resource. The possibilities of influencing the outcome depends to a large degree on their organisational abilities, their connections to the central political authority and, more importantly, on the institutional order governing the distribution (Apostle, *et al.* 1998; Hersoug and Holm, 1999).

We may identify three institutional orders (or paradigms) governing the distribution of access rights to a resource. These are the community, the market and the state (Apostle *et al*, 1998; Hersoug and Holm, 1998). The community governance paradigm is characterised by close inter-personal ties amongst members, multiplex social networks and shared identities (Hersoug and Holm, 1999; Leach *et al.*, 1999). The community paradigm emphasises the equitable distribution of access to resources, and therefore provides solutions in situations where this is absent (Hersoug and Holm, 1998; Hersoug, 1996). The community paradigm tends to favour small, local and traditional participants or user groups. The concepts of local management and local user participation seek to embed fisheries management in local community structures (Pomeroy and Williams, 1994; Jentoft, 1989). This paradigm is therefore difficult to establish where the fishery is exploited by highly capitalised companies comprising heterogeneous owners.

The state governance paradigm is broadly characterised by a top-down, centralised bureaucratic command and control mode of decision-making (Hersoug and Holm, 1998; Pomeroy and Williams, 1994; Apostle *et al*, 1998). This paradigm is recommended in a situation where over-exploitation and the unfair allocation of resources amongst nationals are the main problems of a fishery. The central state is seen, therefore, as an important actor in transforming social relations. In a country such as Zimbabwe, however, politicians do not always remain in control as organised interest groups influence important policy decisions.

The market-based property rights resource management system suggests competition, economic efficiency and rationality (Apostle *et al*, 1998; Hersoug and Holm, 1999). It aims to deal with situations where over-capacity, underdevelopment and the inefficient allocation of resources are the main problems of fisheries management. The tradable licenses, leases and quotas now implemented in many countries seek to redress the inefficiencies of state governance by utilising market mechanisms (Copes, 1996). Some market governance has, however, had inequitable consequences. For example, Individual Transferable Quotas (ITQs) for Icelandic cod have caused control of the industry to shift to a few large operators (McCay and Acheson, 1987), resulting in reductions in the number of firms and jobs. Policies grounded in the market paradigm fail to address issues of social equity, as the goal of economic efficiency often clashes with that of maintaining rural communities (Waters, 1991; Corten, 1996).

Each of the above three institutional orders has the capacity to redistribute access rights to natural resources in somewhat different directions. Under the market paradigm, actors able to exploit the resource using the most efficient strategies will gain control over it. Relying on the community paradigm, however, could undermine the big companies in favour of local communities. The governance structure of the state emphasises control. The process of redistribution, however, depends on the extent to which politicians retain complete control, and are able to resist the influence of lobbying groups and negotiations over policy (Hersoug and Holm, 1999).

This paper seeks to discuss the key policies and statutes in fisheries that influence the redistribution of *kapenta* fishing licenses on the Zimbabwean side of Lake Kariba. The type of and capacity for redistribution, however, is a question of choice of governance paradigm (Hersoug and Holm, 1999). Whatever the choice made, the goal of redistribution involves making trade-offs between the incompatible requirements of the three institutional orders of state, market and community. Each of the three paradigms has had an influence on the effort of redistributing licenses within the *kapenta* fishery. The paper focuses on how licence redistribution has fared given tensions generated by the need to take into account (a) the pressure for retaining the role of the state; (b) the participation of users and (c) the pursuit of economic efficiency through market solutions.

The evidence presented in this paper is based on fieldwork conducted between August 2000 and March 2001. Key informant interviews with purposively selected *kapenta* fishers and resource managers were undertaken. Additional data has also become available through the work of other researchers. These are referred to and discussed in the light of findings.

The *kapenta* fishery's management

The Zimbabwean *kapenta* fishery of Lake Kariba is defined as state property. The state makes decisions concerning access to it, and the level and nature of exploitation. The state employs several measures to manage the resource, including licensing, area closure and area zoning.

Licensing

Non-transferable licenses are issued annually by the state through the Department of National Parks and Wild Life Management (DNPWLM). In

Zimbabwe, a fishing licence is equivalent to a fishing vessel. One company can therefore hold several fishing licences. There are 280 *kapenta* fishing vessels registered to 73 fishing companies, which land up to 18,000 metric tonnes of fish annually (Songore *et al.*, 2000).

Area closures

Shallow areas that are less than twenty metres deep, rivers and river mouths are considered to be breeding areas, and are closed to fishing. Areas near recreational parks and development areas are also closed to fishing to accommodate the other multiple uses of the lake (tourism, sport fishing, and safari operations).

The allocation of fishing basins

Fishing sites were originally allocated near Kariba Town from where most companies continue to operate. With the growth of the industry, new sites were opened in Chalala/Bumi, Sengwa, Binga and Mlibizi basins (see Figure 14). Permits issued contain provisions that limit the basin in which vessels may operate, and licence holders are not permitted to fish outside that basin (see Table 3 for the distribution of fishers by basin/fishing zone). Regrettably, the allocation of fishers per basin is based on unsubstantiated research. It was hoped that by assigning a basin to a given limited and identifiable number of fishers, self-management might be encouraged. Based on some empirical evidence, many scholars argue that only small groups can organise themselves effectively (Ostrom, 1995b; Murphree 1991; Baland and Platteau, 1996), especially if the population is homogeneous. This argument overlooks the fact that communities and their environment are often heterogeneous.

Basin	**White-owned Co.s**		**Black-owned Co.**		**Rural District Councils**		**Co-op.s**	
	Co.	Vessels	Co.	Vessels	Co.	Vessels	Co.	Vessels
Mlibizi	0	0	0	0	0	0	1	2
Binga	6	21	1	2	1	3	5	10
Sengwa	7	20	1	2	1	3	5	13
Bumi	12	64	0	0	0	0	2	4
Sanyati	19	96	6	22	1	4	5	14
Total	44	201	8	26	3	10	18	43

Table 3: Distribution of fishers by basin and race in 2000

There seems to be little reason for restricting fishing in a particular basin when *kapenta* can move anywhere in the lake and the fishing vessels have to search for them. By doing so, they can stray out of the fishing basins and this is unlikely to have any effect on the total fish stock. The zoning regulations have given some operators advantages over others. They have created two types of inequalities. Firstly, the productivity of these basins varies significantly due to variations in zooplankton levels. Total zooplankton densities are always highest in Basin 1 (Mlibizi) followed by Basin 2 (Binga). Basins 3, 4 and 5 (Sengwa, Bumi and Sanyati) always have lower densities than Basin 1 and 2. However, Basin 5 (Sanyati) usually has higher densities than Basins 3 and 4. The plankton density gradient observed from Mlibizi to Bumi is expected of river-fed man-made reservoirs because of the gradient in nutrient concentrations. Sanyati Basin, the most downstream of the five basins, should be the least productive, but this is not so because of the high nutrient inflow from the Sanyati River (Masundire, 1997; Moreau, 1997). Other rivers feeding the lake do not have much effect on pelagic zooplankton because they drain smaller catchments with lower nutrient levels (Masundire, 1997).

Secondly, the zoning approach used did not consider other important issues that affect the economic performance of operators fishing in the different basins. These included

(a) The influence of waves that disrupted fishing, reduced fishing time, and increased wastage. Waves are particularly problematic in Basins 1 and 2.
(b) Distance to the nearest urban centre where operators obtained supplies such as fuel and salt; and
(c) availability of markets for *kapenta*. Fishers in the Sanyati and Binga basins operate from Kariba Town and Binga service centre respectively. The rest operate from remote areas.
(d) Finally, the zoning did not take into account variable ecological conditions and differences in production across the lake. Up to now, regulations linking licences to basins have not been revised. Failure to take into account the variations in resource productivity and the costs of operating in different locales undermines the performance of new entrants who are assigned very remote fishing basins.

The heterogeneity of physical characteristics across the lake causes variations not only in resource productivity, but also in patterns of interaction. Some fishers have contested the creation of these physical boundaries by fishing illegally in other fishers' entitlements. For example, fishers in the Binga basin, where poor water visibility reduces the arc of light attraction, poach in the Sengwa basin. Fishers in less productive basins (Bumi and Chalala) also fish illegally in the Sengwa basin.

Kapenta fishing methods

Kapenta fishing is done at night using light attraction and lift nets. Besides marine engines, a fishing vessel has to be equipped with an electric generator and powerful mercury lights. The dip net is lowered into the water, and the lights then turned on to attract the *kapenta.* When enough fish have gathered above the net it is then hauled, using either a hydraulic or manual winch (Figure 15).

Figure 15: *Kapenta* fishing vessel with rig

The colonial legacy in Zimbabwe

Like many post-colonial states, the government of Zimbabwe inherited a dual socio-economic system based on racial privilege. At independence, the economy was characterised by the inequitable distribution of income and productive assets in favour of a white minority (Government of Zimbabwe, 1981). The *kapenta* fishery was no exception. The initial selection of *kapenta* fishers occurred at the DNPWLM headquarters during white rule. People who had experience in commercial farming (fisheries fell under the Ministry of Agriculture at that time), were eligible for a *kapenta*-fishing licence, provided they could afford it (Marshall, *et al.*, 1982; Hutton, 1991). There was no limitation on the number of licences or fishing vessels that could be held by a fisher, and hence the size of the individual fishing operations that developed varied considerably.

From the outset, therefore, the criteria used to allocate *kapenta* fishing licences favoured white commercial farmers against black peasant farmers. This was accentuated by the fact that whites dominated fisheries management and administration within the DNPWLM.

Historically, there has been a lack of cooperation between white and black communities. Even before the war of liberation, there were separate development initiatives between black and white fishing communities. In Rhodesia, fisheries development fell under two authorities. The Ministry of Agriculture provided extension services to the European controlled fishery, while in tribal areas, extension was the responsibility of the Ministry of Internal Affairs, which had a unit with just one fisheries officer (Rhodesian Herald, 26 August 1975). This resulted in fishing skills and expertise disparities between white and black fishers. Black fishers were restricted to small-scale fisheries, while whites came to dominate the large-scale, multi-million dollar *kapenta* fishing industry.

The initial selection process, therefore, benefited white entrepreneurs the most. The larger population of black people did not benefit, mainly because national education and manpower policies at the time were generally designed to ensure the existence of a cheap and unskilled black labour pool (Murphree *et al.*, 1975). Hence, blacks' entrepreneurial skills were very limited (Bourdillion, *et al.*, 1985). The colonial government depended on the importation of skills mainly from overseas (Murphree *et al.*, 1975), and these policies hindered local skill development, especially amongst blacks. Blacks provided labour to most industries, including the *kapenta* fishing industry where they are still employed as boat crews, fish processors and packers.

In addition, most black people did not have acceptable collateral and could not, therefore, qualify for bank loans to start up *kapenta* fishing. The lack of collateral has an historical explanation. Local blacks have been excluded from ownership of the means of production and from any economic activity that would enable then to make meaningful savings. As a result, only white fishers were licensed to exploit the *kapenta* fishery before independence.

Zimbabwean policies on natural resource distribution

In 1980, the Zimbabwe Government, wishing to redress racial imbalances in resource access, embarked on policies and strategies aimed at promoting the equitable distribution of income, the sustainable use of resources and increasing user participation in the management of natural resources (Government of

Zimbabwe, 1981). To address the inequalities in resource distribution, the government introduced its 'growth with equity' policy in 1981. The policy contains two objectives of direct relevance to natural resource management. The first objective is 'to end imperialist exploitation and achieve a greater and more equitable degree of ownership of natural resources [and to] promote participation in and ownership of, a significant portion of the economy by nationals and the state' (Government of Zimbabwe, 1981: 2). The second objective is '...to conserve our natural resources so that production is sustained, replace renewable resources used, [and to] exploit our natural resources at rates consistent with the needs of present and future generations of Zimbabwe' (Government of Zimbabwe, 1981: 3).

To effectively redistribute fishing permits from established fishers to black fishers, and realise the equitable distribution of the fisheries resources, the Minister of Mines, Environment and Tourism is empowered through Section 82 of the Parks and Wildlife Act of 1975 to revoke fishing permits. In section 83B, the Minister may 'if necessary or desirable to do so in the interests of the preservation, conservation, propagation of any fish within Zimbabwe prohibit any person from fishing absolutely or subject to certain conditions or from possessing fishing gear' (Hutton, 1991: 13). The Act, therefore, concentrates considerable power with the Minister. He may or may not give any reason for the withdrawal of licences. Nor is he obligated to authorise compensation for anyone unduly affected by the termination of a permit.

While appearing to be captious, the sweeping authority of the Minister to issue and revoke licences was presumably intended to give National Parks' wardens power to deal with persistent offenders, such as anglers who exceeded catch restrictions, used inappropriate gear or fished in closed areas. Today, with an industrialised fishery such as the *kapenta*, which employs over 3,000 people, this authority has caused considerable panic amongst established fishers and conflict between the DNPWLM and fishers.

Kapenta fishing was a very lucrative business. Soon after independence, there was pressure on the government from emerging black entrepreneurs to redistribute national natural resources equitably amongst citizens of all races. The government, '...desiring to establish a society founded on socialist democratic and egalitarian principles' (Government of Zimbabwe, 1981: 2) issued twelve licences to emerging co-operatives in 1986. However, despite an increase in the number of fishers, catches did not increase. The Zimbabwean government interpreted this fall in catch per unit of effort as an indication of over-exploitation. The DNPWLM then adopted a precautionary approach to issuing new licences because the size of the biomass was unknown. Licences were issued progressively, and mean fish length and catch size carefully monitored. Only two new licences were issued between 1990 and 1995 and none has been issued since then, while the Department awaits the results of a stock assessment programme. Because of the government's concerns over fish stocks, therefore, equitable distribution could only be achieved through the redistribution of existing licences from large companies (mostly white owned) to small and emerging companies and co-operatives (mostly black-owned). In the process, three types of business organisations emerged within the *kapenta* fishery. These were public ventures (run by local authorities), co-operatives and sole proprietorships owned by both white and black fishers. This development also resulted in two institutions being created along racial lines that mediated on behalf of their respective members on issues regarding licence redistribution. The Indigenous *Kapenta* Producers Associations (IKPA) represented the interests of black fishers, and seeks to lobby for the equitable redistribution of licences. The IKPA has political clout and sympathy derived from government. On the other hand, the *Kapenta* Producers Association (KPA) represents white fishers' interests. Its main mandate, amongst others, is to protect the interests of members during the restructuring process. It is interesting to note that the management of the fishery is not on the agenda of the two associations.

Scientific socialism and co-operative development

Zimbabwe's post-independence policies of scientific socialism tended to view any private business as individualistic, exploitative and parasitic (Maphosa, 1996), a policy environment not conducive to the creation of small private business. Individually owned businesses had considerable difficulties qualifying for government assistance. As a result, there were only eight black-operated companies within the *kapenta* fishing industry at this time. The government's attention and energy was instead directed towards co-operative enterprises, thereby extending socialist and popular democratic participation in the ownership and the management of natural resources.

The government therefore set up training facilities for co-operative members so as to ensure the democratic, orderly and profitable functioning of co-operative enterprises (Government of Zimbabwe, 1981). In 1986, the government facilitated the formation of co-operatives for the commercial fisheries of Lake Kariba. There are now 18 co-operatives within the *kapenta* fishery, which control 15% of the fishing vessels on the lake. Most co-operatives started with initial group savings and sought assistance from government. Government controlled parastatals (for example, the Small Enterprise Development Corporation) and government departments (Ministry of National Affairs Employment Creation and Co-operatives (MNAECC)), advanced soft loans to the co-operatives to enable them to improve their management and operational capacity. However, the cost of setting up co-operatives far exceeded what parastatals and the government were prepared to pay. Therefore, members had to seek financial assistance from international aid organisations. Non-governmental organisations, such as MS Zimbabwe, and government departments provided training in managerial skills. The idea of co-operatives was an attempt to involve indigenous Zimbabweans in the exploitation of the *kapenta* fishery and, in a sense, simulate the community paradigm discussed earlier.

The co-operatives were supposed to survive in the same way companies in business do. They were set up to compete in conditions determined by other companies that had been in business for some time and which had developed highly efficient production techniques and an elaborate understanding of market conditions. Some of these companies had operated during extremely adverse periods, such as under the economic sanctions during the Unilateral Declaration of Independence period, and had developed very good planning and management strategies.

The promotion of co-operatives exposed new fishers to stiff competition with established white fishers. Besides, these established companies were capable of operating in the hardest of circumstances because of the economies of scale acquired over years of fishing. White fishers were not only engaged in fishing and processing, but also in retail marketing, and offered a wide range of products for both local and export markets. Some of these fishers also had facilities to market fresh fish. Wealth inequalities based on differences in skills, experience, better management and a willingness to accept greater risks have persisted over the years. Some co-operatives have gone bankrupt, and others have failed to service loans. There are several reasons why these co-operatives did not perform well. Firstly, there was inexperience, and internal financial abuse within some of these co-operatives. Secondly, most co-operatives operated in remote areas, making it difficult to buy inputs and to market their *kapenta*. Thirdly, the average co-operative operated three fishing vessels, which was insufficient to support co-operative members, service loans and meet all operational costs. Fourthly, the launching of economic structural adjustment programmes in 1991 weakened the government's thrust towards socialism and opened the economy to market forces. The government then withdrew financial support for co-operatives. The withdrawal of parastatals from the provision of credit has been replaced by the private sector. However, probably because of high transaction costs and the inability to enforce contracts with co-operatives (Maphosa, 1996), private financiers are unwilling to provide credit to co-operatives. Finally, the high transaction costs associated with the democratic control and management of business operations probably accounted for much of the poor performance of the co-operatives.

As a result, some co-operatives have stopped operating and now earn an income by leasing out their licences, fishing vessels and premises to wealthy and established white fishers (Moinuddin, 1991). Under this arrangement, the ownership of licences remains with the licensee who then transfers the management of his vessel to more efficient operators. One can therefore operate several vessels that may still be formerly owned by another fisher. This violates the non-transferability of the licences, and the practice has facilitated the emergence of large oligopolistic companies in some basins. The objectives that the co-operatives were intended to achieve no longer guide their activities. This complicates the management of the fishery, especially where resource users are called upon to contribute to this.

There were, however, some co-operatives that did succeed. Most of these were based in Kariba and at Binga, where proximity to input supplies and readily available fish markets probably contributed to their success. There were also some entities run by individual black fishers, which, although displaying some teething problems, are now performing well.

Licence redistribution during the first ten years

Despite efforts to redistribute access rights to the fishery through affirmative action, white-owned companies have continued to increase both in absolute numbers and in the size of their operations since independence in 1980 (Table 4).

Year		Whites	Blacks	Co-op.s	Local authorities	Totals
1979	Fishers	22	0	0	0	22
	Boats	61	0	0	0	61
1980	Fishers	37	0	0	0	37
	Boats	132	0	0	0	132
1985	Fishers	33	0	0	0	33
	Boats	156	0	0	0	156
1990	Fishers	39	5	5	3	52
	Boats	199	1	12	2	214
1995-2000	Fishers	44	8	18	3	73
	Boats	201	26	43	10	280

Table 4: Changes in the categories of fishers on Lake Kariba (Zimbabwe). Source: Lake Kariba Fisheries Research Institute Database 2000

In the early 1980s, a number of factors combined to increase the importance of the Zimbabwean commercial fishing industry. For example, during the drought years of 1982, beef supplies diminished and organisations such as the World Food Programme purchased large quantities of *kapenta* for famine relief in the rural areas (ZZSFP, 1989). The distribution of dried *kapenta* to many parts of the country where fish was not a traditional part of diet resulted in a much wider acceptance of the product (ZZSFP, 1989). Increases in meat prices have further increased demand for *kapenta*. In response to this, the government gradually issued more licences to new and existing companies irrespective of race. For the first ten years after independence, the government was more concerned with the national food security than correcting racial imbalances. The drought of 1982 derailed the government's initiatives towards equitable distribution.

For most fishing companies, fishing is the main livelihood source. For this reason, they have invested a lot of money and time in the fishery. The redistribution of licences is politically motivated, and there is no biological or economic justification for it. With very little knowledge regarding the status of the *kapenta,* managers become very conservationist and set arbitrary harvesting levels far below resource threatening levels (ZZSFP, 1997b). In this situation, the chronically unsatisfied white fishers defend their access rights and pressurise resource managers to implement liberal harvest levels instead of redistributing their licences. They argue that many more black fishers can be recruited into the fishery without taking licences from the whites and, at the same time, without putting the fishery at risk of collapse. The economic and political power that the established white-owned companies have acquired over the years has allowed them to maintain their positions. Among the white operators, there are very few who do not know at least one cabinet minister personally. Some ministers used to come to buy *kapenta* in bulk at producer price. White fishers use this acquaintance to protect their access rights. The state governance paradigm does not, therefore, guarantee equitable resource redistribution amongst nationals, especially where there is skewed wealth distribution. Lobbying and negotiations can influence state governance and it is the wealthier group with greater economic and political power that succeeds in maintaining or even strengthening its position.

The willing seller, willing buyer principle

With pressure from Indigenous *Kapenta* Producers Association (IKPA) and a desire to please the electorate in preparation for 1990 general elections, the government sought to speed up the redistribution of access rights into the fishery. To this end, both the DNPWLM and fishers agreed that the maximum ceiling for permits per entity should be nine and that the minimum should be four. This was designed to benefit both small companies owned by blacks and emerging black entrepreneurs. Established companies, in turn, were supposed to dispose of their licence/fishing vessels on a willing seller willing buyer basis, an arrangement designed to simulate pure market processes. The upper and lower licence limits were, however, flawed. As one operator put it:

> 'It is pointless to have variations in the number of licences by company because the costs of running a company are the same. You need the same costs for a manager, an accountant and a welder irrespective of company size.'

So as to avoid a situation in which established companies with good credit facilities were able to buy up smaller and more vulnerable companies, the DNPWLM was supposed to identify suitable (black) buyers for these companies. The willing seller willing buyer principle was expected to apply even after redistribution had occurred. The price of a *kapenta* fishing vessel was, however, far too high for black entrepreneurs. This frustrated efforts to see the redistribution process through, and the licence redistribution programme did not yield its intended results.

The number of white-owned companies continued to increase between 1990 and 1995 as shown in Table 4. Fears of alienating international donors, investors and powerful private capital, largely in the hands of whites, have forced the government to proceed with caution in its efforts to address racial imbalances. Redressing the inequitable distribution of access rights to natural resources in Zimbabwe has degenerated into a conflict between the state and the Western superpowers. The market solution (guided by the state's administrative machinery) was used without adequate credit facilities and sound capacity-building initiatives. This has led to the establishment of fishing operations operating very few fishing vessels that fail to operate above the break-even point.

The sliding scale

Following the failure of the willing seller to release enough licences on to the market for the willing buyer to purchase, the IKPA made representations to government seeking equitable licence redistribution. The government introduced economic disincentives for established companies to hold many licences. In May 1995, the Minister withdrew and terminated all commercial fishing permits pending a decision regarding the conditions of reissue (Herald, 13 June 1997). In 1996, the DNPWLM forwarded a second notice to operators informing them that as of January 1996 there was to be a reduction in the number of permits held by each operator. The notice also advised operators that a sliding scale for licence fees was to be introduced. The sliding scale was designed to increase production costs for those with more than two fishing licences/vessels. This development was, however, contested in law courts by the *Kapenta* Producers Association, representing the interests of white operators. In 1997, the sliding scale was ruled invalid (Zimbabwe Independent, 1997), and has not been effective in delivering access rights to disadvantaged black people. This has complicated the collaboration between the fishers and the state in the management of the fishery. At the same time, the majority of established fishers live in a state of considerable uncertainty concerning their futures.

Indigenisation

In 1998, the government, under pressure from black businesses collapsing on account of globalisation, introduced an 'indigenisation' policy. The policy seeks to reduce poverty amongst the majority of Zimbabweans. Two of the objectives of indigenisation, which have a direct impact on the *kapenta* fishery, are, firstly 'to democratise the ownership of the economy so as to eliminate racial differences arising from economic disparities', and secondly, to develop entrepreneurship and economic management skills through training (State Enterprises and Indigenisation Department, 1998: 12). The policy created favourable conditions for the advancement and empowerment of disadvantaged groups in Zimbabwe in order to bring about equity amongst the different groups.

Under this policy, the natural resources endowment of a province determined the areas of indigenisation (State Enterprises and Indigenisation Department, 1999). In the case of Kariba, the resources that formed the basis for indigenisation were the *kapenta* fishery and the gill net fishery. There were, however, questions within the *kapenta* fishing industry regarding the definition of the term 'indigenous'. Although some operators argued that it implied the 'black' race, objections were raised by established operators who argued that the term applied to all those who were born and brought up in Zimbabwe. Indigenisation has not achieved much in terms of black advancement, possibly because the policy is still at its introductory stage within the *kapenta* fishing industry.

Conclusion

The *kapenta* fishery has demonstrated that institutional arrangements have very little relevance to guaranteeing the efficient management of a fishery and the equitable distribution of access rights to it as long as there is inequality amongst resource users. Management regulations that are designed to protect the resource against overexploitation and degradation may actually lead to the polarisation of its users, and become a source of conflict between them.

Restructuring of the *kapenta* fishing industry is based on affirmative action and the need to reverse discriminatory practices, and not on technical merit. No considerations were given to the likely disruption this restructuring would cause to investment patterns and economies of scale acquired over years. The process within the *kapenta* fishery focused on the equitable distribution of access rights, and therefore failed to change inequalities that were created within the credit, input and output markets. Information on markets and technological development remain skewed in favour of established companies. As a result, inequalities based on the appropriation of skills, initial endowment, access to credit and markets persists within the industry.

As has been demonstrated in this study, inequalities distribute incentives in different directions resulting in antagonistic relations amongst resource appropriators and between appropriators and the state. Baland and Plateau (1999) have argued that in a regulative setting, inequalities tend to reduce the acceptability of available regulative schemes by the most vulnerable group, making collective action in management very difficult. Though this may be the case within the *kapenta* fishery, inequalities have brought together fishers (along racial lines) to advance their collective best interests.

6

The Outcome of a Co-Managerial Arrangement in an Inland Fishery: The Case of Lake Kariba (Zambia)

Isaac Malasha

In most Southern African countries, Community-Based Resource Management (CBRM) initiatives have been instituted to manage a variety of natural resources such as forests, wildlife and fisheries. In Zambia, there is the Administrative Management Design for Game Management Areas (ADMADE), the Living in a Finite Environment (LIFE) in Namibia and the Communal Areas Management Programme for Indigenous Resources (CAMPFIRE) in Zimbabwe. These initiatives seek to confer some managerial responsibilities on local resource users as opposed to central and government-driven management regimes. Within the fisheries sector, 'co-management' has emerged as another variant of the CBRM approaches. Co-management systems have emerged as a partnership arrangement using fishermen's knowledge of the resource and their organisational capacities complemented by the ability of government to provide enabling legislation, enforcement and conflict resolution (Pomeroy and Berkes, 1997). The partnerships between government and local resource users can range from those in which fishers are merely consulted by government to those in which fishermen implement and enforce fishing regulations.

The advantage of co-managerial arrangements is that they are more economical in administration and enforcement of fishing regulations than centralised systems (Pomeroy, 1994). Co-managerial arrangements also provide a sense of ownership over a fishery, which increases the potential for long-term sustainability. Jentoft (1989) has further observed that while the co-management arrangements are not a panacea for solving all the problems of fisheries management, when benefits and costs are taken into account, co-management must be considered a viable option compared to other management alternatives. There is also a greater likelihood that access to the resource and its equitable distribution will occur under co-managerial arrangements than under a government controlled management regime (McCay, 1993). This is usually based on the understanding that local actors are more familiar with local conditions than governments or their agents.

Studies of co-management arrangements have shown that these systems are best achieved where users already have a cohesive social system based on kinship, ethnicity or homogeneous gear type (Pinkerton, 1989: 28). This element of homogeneity creates pre-conditions for fishers to co-operate in the management of a fishery.

The problem with such an understanding of co-management is that it simplifies complex human behaviour to linear and predictable interactions in the access to, and utilisation of, natural resources such as an inland fishery. The emphasis on homogeneity either in terms of ethnicity or type of gear used as a positive factor in successful co-management arrangements ignores the different power relations among fishermen and other interested parties, which may actually determine how a fishery is to be accessed and utilised. Furthermore, the emphasis on co-management arrangements as leading to the equitable distribution of benefits from a resource does not account for the multiple uses of the fishery and the numerous livelihood strategies of artisanal fishermen. These varied uses and strategies may actually lead to increases in the unfair distribution of the resource and a rise in conflicts amongst different users. As examples from Lake Kariba will show, the introduction of co-management was contested by various groups of local actors in the fishery, such that the outcomes led to conflicts and the marginalisation of artisanal fishermen who, initially, had been expected to benefit from the new managerial regime.

Lake Kariba

The primary use of Lake Kariba, located on the Zambia/Zimbabwe border, is to generate hydroelectricity for the two countries' industrial sectors.[13] The dam wall on the Zambezi River was sealed in the late 1950s creating a reservoir covering a water surface area of 5,364 km². It has a shoreline length of 2,164 km. It is about 300 km long and about 40 km wide at its widest point. Approximately 55% of the lake's area lies within Zimbabwe, and the remainder in Zambia. Lake Kariba's main tributary is the Zambezi River, which contributes about 70% of the water inflow.

The lake has two distinct fisheries: the artisanal and the semi-industrial sectors. The latter sector exploits a single fish species, the '*kapenta*' (*Limnothrissa miodon*). The *kapenta* was introduced to Lake Kariba in 1967/8 from Lake Tanganyika, to occupy a vacant ecosystem niche that had developed in the lake's pelagic waters. It is harvested by the use of mechanised fishing rigs and dip-nets. The sector, therefore, is a preserve of individuals

---

[13] Unless specifically mentioned, Lake Kariba in this paper will refer to the part of the lake lying in Zambia.

and companies with the necessary capital to invest in these rigs. Approximately 8,000 t of *kapenta* is harvested annually (Chitembure *et al*, 1998)

The inshore fishery of Lake Kariba consists of those parts of the lake which are not more than twenty metres in depth (Chitembure *et al.* 1999; Sanyanga, 1996). In this part of the lake, the fish species evolved in riverine habitats and can only inhabit the marginal areas of the lake. About 50 species of fish occupy these waters, of which artisanal fishermen commercially exploit close to fifteen. Artisanal fishing is conducted with multi or monofilament gill nets, set by largely non-motorised vessels, ranging from metal and fibreglass boats to dugout canoes. Engine-powered fishing boats are rare, and those that exist are normally used for transportation and not fishing. The inshore fishery produces about 2,000 t of fish in a year (Chitembure *et al*, 1999). Nearly 80% of the catch from this sector is sold, the remainder being absorbed by spoilage and home consumption (Hachongela *et al.*, 1998).

While fishing may be the main source of income for Lake Kariba's artisanal fishermen, about half of these fishermen are also involved in other economic activities such as agriculture, livestock keeping and fish marketing (Walter, 1988). Whenever possible, activities such as agriculture and livestock keeping are conducted near the lakeshore.

The artisanal fishermen

Before Lake Kariba was created, the local Tonga people, whose land was submerged by the rising waters, were a predominantly farming people. Therefore, when artisanal fishing started in the late 1950s the Gwembe Tonga Native Authority prevailed upon the government of the day to allow only the Tonga into the fishery for the first ten years. This was done because most of the local people would have to take up fishing as there was going to be inadequate land to continue with their agricultural activities. This ethnically-based entry system was intended to create opportunities for them to take up commercial fishing without competition from other groups with already well-developed fishing traditions. Enforcement of this entry system was made easier by the various pieces of colonial legislation that restricted the movements of local people.

Initially, the Tonga took to fishing enthusiastically. In 1962, about 3,000 t of fish were caught and the following year catches rose to about 4,000 t (Bourdillon *et al.*, 1985). From 1965, however, catches began to go down and in 1968, only 800 t of fish were caught (Bourdillon *et al.*, 1985). The decline in catches has been attributed to the loss of initial nutrients through the hydro-electric turbines (Karenge, 1992). As catches began to decline, most of the Tonga fishermen left the fishery and invested their savings in commercial agriculture and livestock (Colson, 1971; Scudder, 1960). Those who remained in the fishery became largely part-time fishermen spending most of their time tending to crops and livestock.

After Zambia's independence in 1964, the ethnically based entry restrictions into the fishery were abolished, along with legislation that restricted the movement of people, especially in the rural areas. Immigrant fishermen were subsequently allowed to enter Lake Kariba. Presently, about half the artisanal fishing population is Tonga and the remainder are immigrant fishermen (Walter, 1989). The Bemba and Lozi, who have a long fishing tradition in their home areas, constitute one of the largest immigrant groups in the Lake Kariba fishery. Apart from the abolition of restrictions, the entry of immigrant fishermen into the fishery seems to have also been largely prompted by external factors. As employment opportunities have declined in the formal sectors of the economy, immigration to the fishery has increased. Between 1990 and 1993, the number of fishermen on Lake Kariba grew from 1,733 to 2,283, which may be attributed to employment declines in the formal sector over the same period (Chitembure *et al.* 1999).

Kapenta operators

Kapenta fishing on Lake Kariba started after 1980. Due to the high cost involved in obtaining fishing rigs, processing plants, as well as having adequate capital to meet the cost of fuel and spare parts, the harvesting and marketing of *kapenta* is much more capital intensive than artisanal fishing. Consequently, kapenta operators tend to be involved in running other economic activities such as commercial agriculture, crocodile farming, and construction. Activities such as crocodile farming are directly related to *kapenta* production as the crocodiles are fed on kapenta. A number of *kapenta* operators have also made investments in facilities for tourists. Most of these operators are members of the *Kapenta* Fishermen's Association (KFA).

The linkage between the artisanal and *kapenta* sectors is mainly in the recruitment of labour. Most of the labour employed in the *kapenta* industry is drawn from the artisanal sector and from Tonga villages on the lakeshore. Within this sector, one indirect cost incurred is the theft of *kapenta* by workers who operate fishing rigs. Some *kapenta* is stolen in processing plants, but most of it is stolen on the rigs during the night and sold to illegal traders before owners account for it. Theft from the rigs is made possible because *kapenta* crew members operate far from shore where supervision is difficult and the potential for disposing of some of the catch very high. Traders, who may pretend to be buying fish from artisanal fishermen, especially those who fish from the numerous islands in the lake, sometimes buy such stolen *kapenta*.

It is estimated that between 30 and 60% of the *kapenta* is stolen either in the processing plants or on the rigs (Boe, 1998; Lupikisha[14] pers. com.).

Management of the fishery

From the time artisanal fishing commenced up to Zambia's independence in 1964, the fishery was managed by an institution known as the Gwembe Tonga Native Authority (GTNA) with the technical assistance of the then Department of Game and Tsetse Control (DGTC). The GTNA was a colonial institution, which comprised the local Tonga chiefs, their councillors, and the District Commissioner.

Following the abolition of the GTNA after 1964, the role of Tonga chiefs in the management of the fishery declined. Management was now assumed by the Department of Fisheries (DoF), which took over the responsibilities of the DGTC. The DoF was assisted by the local authority in the provision of infrastructure, such as fish markets and roads to the fishing camps. The DoF has managed the fisheries under the provisions of the Fisheries Act, which defines the manner in which fishing is to be conducted and the sanctions to be applied to violations. The DoF was, however, designed more as an extension vehicle, imparting messages on good fishing practices. It was not designed as an enforcement agency, a role further limited by budgetary constraints. It has not, therefore, been able to effectively enforce the provisions of the Fisheries Act, especially with respect to the licensing of fishermen. Consequently, the DoF was unable to monitor and control the entry of immigrant fishermen into the fishery when the ethnically-based restrictions were abolished after 1964.

Concomitant with this uncontrolled entry of immigrant fishermen was an increase in criminal activities in the area, especially after 1980. According to a local Tonga Chief, Sinazongwe, most of these criminals were posing as artisanal fishermen and taking advantage of the scattered nature of fishing settlements to engage in cross-border smuggling and poaching wildlife in neighbouring Zimbabwe.

The introduction of co-management to the Lake Kariba fishery

In 1989, the DoF obtained donor funding to coordinate research and development activities in the fishery. It was under the auspices of this Zambia/ Zimbabwe SADC Fisheries Project (hereafter 'the Project') that co-management was conceptualised and implemented. The rationale for introducing this new management system was to control the entry of fishermen

[14] Pers. comm. with Justin Lupikisha former manager of a *kapenta* company (20/10/2000).

into the fishery, to control fishing settlements and to enforce fishing regulations that were not being observed by the scattered and nomadic fishermen. It was argued that the uncontrolled entry of immigrant fishermen into the fishery, and the subsequent increases in the size of fishing settlements along the lakeshore, was encouraged by the lack of property rights for existing fishermen[15], and gave rise to the selfish and destructive exploitation of the resource (Chipungu and Moinuddin, 1994). The evidence of this destructive exploitation was given as 'indications of a decline in fish size and an increase in fishing in breeding areas and catching of juveniles and use of prohibited fishing methods such as fish-driving.' (Chipungu and Moinuddin, 1994: 3). Therefore, the broad objective of the co-management arrangement was to improve security of tenure for artisanal fishermen through the conference of property rights over particular fishing grounds.

This Community-based Resource Management (CBRM) approach was analogous to the ADMADE management system that was already being implemented in the country's wildlife sector. The new co-management plan for the Lake Kariba fishery was aimed at establishing '...a community-centred resource management regime with responsibilities on the shoulders of the direct beneficiaries which is based on the assumption that the most efficient and cost-effective system of regulation is collective self-interest in sustainability' (Chipungu and Moinuddin, 1994: 4).

In early 1994, the Project hired consultants to obtain views on the new management arrangement from various actors in the fishery. Views were obtained from Tonga chiefs and their headmen, local authority representatives, DoF officials, artisanal fishermen and representatives of Non-Governmental Organisations (NGOs) operating in the area. Upon compilation, the Project sponsored a workshop to discuss the consultants' report. Fifty-six participants attended this workshop, most of whom were Project staff, chiefs, their headmen and local authority officials. At the time, there were about 2,000 fishermen in the fishery, who were represented by just seven participants (Chipungu and Moinnudin, 1994).

It was at this workshop that a framework for a new management plan was adopted. For administrative purposes, the workshop participants agreed to divide the fishery into four zones. Each zone was to be an area of the lake and the hinterland falling under the jurisdiction of a particular Tonga chief. A Zonal Management Committee (ZMC) was to be in charge of each zone. Membership of the ZMCs was to comprise the local Tonga chief, a

---

[15] At the time, there were about two hundred and fifty-six permanent and temporary fishing villages and settlements along the entire shoreline.

representative of the local authority, a member of the KFA, four artisanal fishermen from the zone and '...businessmen with active and well-established businesses along the shoreline,' (Chipungu and Moinuddin, 1994: 7). The ZMCs were to be responsible for monitoring the implementation of fishing regulations, drawing-up development plans and managing a revolving fund.

Below the ZMCs, the workshop agreed to the creation of an Integrated Village Management Committee (IVMC) in each fishing village. The IVMCs comprised a chairperson elected by the artisanal fishermen from the fishing village, the local Tonga headman, three elected fishermen, a DoF representative and an honorary fish scout appointed by the DoF. Collective management rights were to be delegated to IVMCs. The IVMCs were to be responsible for recommending new entrants into the fishery and were to assist with the enforcement of fishing regulations. The IVMCs were also responsible for imposing sanctions on those who were caught violating the fishing regulations in fishing grounds under their management.

The ZMCs and IVMCs were to be funded through licence fees collected by the DoF, fish-levies collected by the local authority, contributions made by members of the KFA and an initial DoF contribution through the Project. The ZMCs held their meetings at the DoF offices while the IVMCs met in the fishing villages.

The workshop also agreed that fishermen would have to operate from designated fishing villages along the lakeshore. Out of the existing 265 permanent and temporal fishing villages and settlements along the lakeshore and islands, only 50 were to be officially recognised. Consequently, fishing settlements on islands and other places were to be abolished. These new fishing villages were to be identified and demarcated by DoF officials, local Tonga chiefs and staff from the local authority. Seven fishing villages were demarcated in Zone Two where data for this paper was collected.

The movement of fishermen into the designated fishing villages presented different opportunities to the various actors involved in the new management plan. The DoF was supportive of confining fishermen to designated fishing villages, as this would improve the monitoring of fishing regulations. It would also improve the licensing of fishermen, and hence control entry into the fishery. The creation of these fishing villages was also seen as making it easier for the local authority to collect revenue from the fishermen. It also made it easier for the local authority and NGOs to provide social services such as schools, hospitals and other amenities. Fish levies were an important source of revenue for the local authority whose funds from central government were declining. The *kapenta* operators saw the new plan as contributing to the reduction of *kapenta* theft as artisanal fishermen, who were said to be colluding with illegal *kapenta* traders, would be confined to specific fishing villages and fishing grounds. It also provided these operators with an opportunity to expand their business ventures into areas from which artisanal fishermen had been removed.

Once the rationale and framework for the new management plan had been agreed upon at the workshop, the implementation phase was set in motion. This initially began in mid-1994, when Tonga chiefs and headmen, accompanied by the DoF and local authority officials, toured all fishing camps and islands to explain the modalities of the new management plan. Fishermen were informed that they would have to move into the designated fishing villages. During these tours, the Tonga chiefs accused immigrant fishermen of not identifying with the local people and their traditional leadership. The immigrants were also accused of engaging in criminal activities. The local Tonga chiefs believed that immigrant fishermen deliberately engaged in destructive and illegal fishing methods so as to maximise their catches before going back to their home areas once the Lake Kariba fishery collapsed. Settlement on islands, which had provided good fishing grounds, was banned and all fishermen had to reside in the designated fishing villages on the mainland.

The role of the *Kapenta* Fishermen's Association

When members of the *Kapenta* Fishermen's Association (KFA) realised the opportunities to be derived from the new management plan, they decided to fully support it. The reduction of fishing villages along the lakeshore and on the islands, and the confinement of artisanal fishermen into designated fishing villages, meant that it would now be easier to monitor the activities of artisanal fishermen and the fish traders. The KFA believed that these measures would reduce and control the *kapenta* theft in the fishery. They therefore made financial and material contributions to the DoF and the ZMCs to assist with the resettlement of fishermen from the islands to designated fishing villages. In conjunction with law enforcement agencies and the IVMCs, the KFA took it upon itself to erect barriers on major roads leaving the fishery to ensure that only legitimately acquired *kapenta* left the fishery. These measures were supported by the IVMCs because about 60% of the levies from legitimate *kapenta* sales were supposed to be channelled to these committees.

In addition, the KFA advocated for the creation of a patrol system to contain *kapenta* theft and other illegal activity. In particular, the members of the KFA wanted to prevent artisanal fishermen from going back to some of the islands, where some members of the KFA had shown an interest in setting up tourist facilities. Operating under the auspices of the ZMCs, and in agreement with the Tonga chiefs and the local authority, the KFA funded the formation of a Marine Patrol Unit (MPU). The unit was equipped with vessels and gave itself powers to arrest any artisanal fishermen or trader engaged in illegal *kapenta* trading or who attempted to go back to closed down fishing camps. At its inception, the MPU had a full-time manager and employed a number of retired policemen and army personnel.

Access to, and control of, the islands

The islands of Lake Kariba became an important part of the inshore fishery when productivity began to decline and stabilise in the mid-1960s. Mainly immigrant fishermen trying to avoid conflicts over land with local Tonga people on the mainland inhabited these islands. They are sparsely populated, and the fishing grounds around them were, therefore, rich. With the introduction of co-management and the eviction of fishermen, the islands became a potential source of income. As soon as artisanal fishermen were evicted from the islands, some members of the KFA proposed to the local authority and the local Tonga chiefs to lease out the islands to set up tourist facilities, such as game viewing and angling. The local authority and the chiefs were supportive of these proposals because they would bring 'development' to the area beyond that brought by artisanal fishing alone.[16] Once the islands had been leased to some members of the KFA, the MPU was mobilised to prevent the setting of gill nets by artisanal fishermen near these islands. Members of the unit arrested artisanal fishermen found fishing on or near the islands, and confiscated their fishing equipment.

Access to, and control of, fiscal revenue

One problem that the local authority faced was that fishing settlements were scattered along the entire lakeshore, which made it difficult to collect any kind of revenue from fishermen. Faced with reduced funding from central government, the local authority was compelled to rely on its own resources to obtain revenue for its activities. The large number of fishermen and the volume of fish taken from the fishery represented an enormous potential for increased revenue. The local authority could increase its revenue base if some of the islands could be leased to investors such as tour operators who could pay higher levies.

In order to legitimise the eviction of fishermen from islands and other closed fishing places, the local authority passed a number of by-laws enforcing the resettlement of all licensed fishermen into designated fishing villages. These by-laws banned the unauthorised utilisation of, or residence on, any of the

---

16 Interview with Chief Sinazongwe, 10/9/1998.

islands under its jurisdiction.[17] The local authority, in concurrence with the chiefs who are said to be the traditional custodians of the land, then proceeded to lease half of the 33 islands in the lake to some members of the KFA for establishing tourist facilities. The original lease period of five years was rejected by the KFA because they would not be able to realise any profits within such as short period. The lease period was eventually increased to 25 years (ZZSFP, 1997a). Unleased islands would still be available to artisanal fishermen upon obtaining a permit from the local authority. This, however, did not authorise the artisanal fishermen to stay longer than a specified period or to set up permanent dwellings, as had been the case before the introduction of the plan.

Despite allowing fishermen to utilise some of the unleased islands, the MPU intensified its operations against some artisanal fishermen who had started going back to the islands or who were setting their nets near leased islands. The KFA supported the activities of the unit so that a good stock of angling fish species was maintained. The unit was also employed to prevent the poaching of game that had been introduced to some of the islands. In one incident, the MPU shot at, and injured, an artisanal fisherman setting his nets near a leased island. The incident caused a conflict between members of the KFA and the artisanal fishermen. The local Member of Parliament (MP) called for an urgent meeting of the ZMC to discuss the conduct of the KFA towards the islands in general (ZZSFP, 1997a). The MP was probably entirely unaware of the activities of the MPU. He may, however, have decided to side with the artisanal fishermen at the expense of the KFA because of the political potential that the artisanal fishermen represented.

At the meeting, it was resolved that only portions of not more than 15 acres of island area could be allocated to members of the KFA. The KFA was accused of creating an illegal police unit to patrol the lake and was asked to disband it (Department of Fisheries, 1995). The DoF and the local authority were also asked to devise new ways that would allow the artisanal fishers to return to islands under certain conditions.

Control of immigrant fishermen

The movement of mostly immigrant fishermen from one part of the fishery to another in search of better fishing grounds had contributed to the proliferation of fishing settlements along the shoreline and the islands. Such movements were also prompted by immigrant fishermen trying to avoid hostilities with local Tonga people over land. When the new management plan consequently

[17] Local authorities are mandated under the Local Authority Act to regulate the collection of levies or creation of new settlements in areas under their jurisdiction.

called for the fishermen to settle permanently in designated fishing villages, it was mostly these immigrant fishermen who were affected. As mentioned earlier, the Tonga chiefs had little sympathy for immigrant fishermen.

In July 1994, the chiefs, headmen and the local authority gave the artisanal fishermen three months to complete their business in the fishing camps that were to be closed down and to move to the designated fishing villages. One group of fishermen was in favour of the management plan and agreed to move. There were, however a small number of fishermen within the group who, while accepting the plan, were sceptical of it. According to informants, this group consisted of those who, in the past, had seen several unsuccessful interventions in the artisanal sector by various government agencies and perceived this exercise to be similar. There was a third group of fishermen who completely resisted the exercise, and some of these hired vehicles to transport themselves and their gear away to other fisheries, such as the Kafue Flats. It is probable that the reduction of fishermen in the fishery from 2,283 in 1993 to 1,355 in 1995 is a result of the implementation of the new management plan (Chitembure, 1995:4).

In December 1994, once the deadline for moving into the designated fishing villages had passed, the Zambia Police, the Marine Patrol Unit and the DoF raided the islands to evict any remaining fishermen. All dwelling huts found on the islands were set ablaze and any remaining residents ordered to leave the islands immediately (Pearce, 1995).

In an effort to convince the artisanal fishermen to remain in the designated fishing villages, some members of the KFA established retail businesses, donated funds to the ZMC to improve their facilities, and offered fishing gear, such as nets, on concessionary terms (ZZSFP, 1995). Artisanal fishermen were also given boats at subsidised rates for as long as they were registered with the IVMCs. Members of the KFA also agreed to continue assisting the ZMC and the IVMCs with finance and material help. By December 1996, however, more than 50% of issued loans had not been returned and were considered bad debts.[18]

Control of arable land

Since the entry of immigrants into the fishery in the mid-1960s, some have been assimilated into the local communities, and consequently, their livelihood strategies tend to mirror those of the Tonga. They own livestock and have access to land for agricultural purposes under different tenure arrangements with individual Tonga. The majority of immigrant fishermen, however, tend

[18] Interview with chairman of the KFA (February, 1997).

to be full-time fishermen with little or no livestock and without access to agricultural land. It is this category of fishermen that tended to move from one part of the fishery to the other in search of high catches. This migration was made easier because the fishermen were tied to neither households nor livestock locally.

This group of nomadic fishermen was heavily affected by the requirement that they move into designated fishing villages, where registers of all fishermen and their gear were kept. For the Tonga artisanal fishermen, however it was the question of access to agricultural land that proved to be a major problem in the new management plan. Although the designated fishing villages were sited with the agreement of the local chiefs, Tonga fishermen refused to move into them because, they argued, this would severely restrict their farming operations. They said that they were not full-time fishermen and saw no reason why they should stay in fishing villages meant for fishermen.

The question of identity seems to have been the major factor that prevented the local Tonga from moving into these designated fishing villages. According to most of the Tonga fishermen interviewed, the designated fishing camps were meant for immigrant fishermen who did not have homes in the area. They claimed that they had legitimate homes from which they conducted their operations, unlike the immigrant fishermen who were visitors to the area. The Tonga felt that moving into these fishing villages would mean giving up their identity as owners of the area. At the height of this conflict, some Tonga fishermen/farmers advocated for the complete removal of the designated fishing villages. This suggestion was not, however, implemented as agreement was reached allowing the Tonga to remain in their farming villages.

The majority of the immigrant fishermen who moved into the designated fishing villages have been unable to obtain land for cultivation. The local Tonga argue that they cannot give immigrants access to land because 'the immigrant fishermen are here for the water and not for land.' One way in which the Tonga prevent the immigrants from gaining access to land is to deny them the right to bury their dead near designated fishing villages. Immigrant fishermen are forced to bury their dead in a cemetery run by the local authority. The local Tonga people interviewed said that the immigrants should have their own graves near the fishing camps or in the local authority cemetery because allowing them the right to bury their dead near Tonga villages might, in future, give them a basis to claim Tonga land.

Access to, and control of, assistance from Non-Governmental Organisations

The creation of institutions, such as the ZMCs and the IVMCs, presented an opportunity for immigrant fishermen to obtain access to resources. Firstly, the immigrant fishermen, who represented a majority in the designated fishing villages, began to dominate elected posts in the IVMCs. The few local Tonga fishermen who had moved into these fishing villages were marginalised from the IVMCs and tended not to be involved in their deliberations. Immigrant fishermen began to use the IVMCs to control access to resources being given to the fishing villages, particularly assistance provided by Non-governmental Organisations (NGOs). NGOs began to assist fishermen in the provision of health services, schools, the creation of women's clubs and other social services. The provision of these services to immigrant fishermen caused other conflicts between the local Tonga people and the NGOs concerned. The NGOs were accused of siding with immigrant fishermen instead of providing assistance to the Tonga owners of the area.

The effect of the co-management plans on the local people

At its inception, the plan was initially seen as solving problems in the inshore fishery only. Consequently, at the time of conceptualising the plan, the only Tonga consulted were either fishermen or members of the local leadership entourage of chiefs and headmen. Implicitly, it would appear that the architects of the plan believed that the Tonga chiefs and headmen would adequately represent the interests of other Tonga people.

This does not, however, seem to have been the case. According to local Tonga people, who are neither fishermen nor members of the leadership, the presence of immigrant fishermen near their villages greatly affected their lives. This disruption was felt more when it came to the question of land. Some of the designated fishing villages were located on land that the local Tonga people used for gardening. Because Lake Kariba had been created, land shortages had occurred in the area. The colonial Gwembe Tonga Native Authority had therefore decreed that the Tonga people would be allowed to engage in agricultural activities wherever arable land was available. One such agricultural practice is known as '*nchelela*' and utilises the moisture retained by the soil as the lake recedes at the start of the dry season. This method ensures the availability of green vegetables throughout the year in an otherwise arid region. Although these types of garden can lie fallow for a number of years, it does not mean that they cannot be reclaimed when the need arises. When the DoF and local chiefs, under the auspices of the ZMC, began to demarcate the designated fishing villages, some of the land on which these new villages were sited was in fact fallow *nchelela* gardens. By establishing permanent fishing villages on such pieces of land, the amount of land available for *nchelela* gardening was drastically reduced. In turn, this affected the local people's household food economy and became a major source of conflict between the local people and the largely immigrant fishermen. Local Tonga people felt that the immigrant fishermen should have been left to operate from islands because 'they are just bringing us problems here.'

Another problem concerning land is that it is simply not available. The unavailability of land was already a problem before the introduction of the co-management plan. As Scudder and Habarad (1991) have shown, emigration by the local Tonga people, which started long before the creation of Lake Kariba, was the result of acute land shortages in the area. The creation of the lake merely accentuated a problem already exacerbated by increased population. The frustration of the local people towards the management plan was aptly captured when one local Tonga farmer observed that:

> 'We are tired of making way for foreign people who come to take our land. Where do they think we shall go, because this is our land? Our children are now finding it difficult to get jobs in the towns and they are coming back to the land. Where are we going to get the land if we give it to anyone including those who come to fish?'[19]

Access to, and control of, co-management plans

It was only after the new management plan had been launched that DoF realised that problems in the fishery extended beyond the need to control illegal fishing methods to include various other resources. While the plan shows that there was inadequate consultation in its conceptualisation, it also shows how the various actors tried to use it to meet their various interests, which, at times, were at variance with the initial objectives.

The evictions from the islands proved to be a source of conflict in many ways. Artisanal fishermen accused the DoF of evicting them from islands to make room for members of the KFA. This became a problem when fishermen were denied access to some of the islands not leased to the KFA. Due to their financial ability, the KFA was able to use the plan to try to protect and expand their business interests in the fishery.

External factors also served to impede the DoF's plans. In 1993, the government launched the Public Sector Reform Programme, which aimed at reducing the size of the civil service by 25% in line with the structural

---

[19] Interview with a Tonga farmer, John Siacheeye, 21/6/99.

adjustment reforms advocated by the World Bank and the IMF. This programme was put on hold in 1996 but re-started in 1997. The DoF was reduced to a small section within the Ministry of Agriculture, Food and Fisheries as a direct result of these reforms. Fisheries extension staff reported to a director (a specialist in crop extension) in charge of all the ministry's extension activities, while Fisheries Research staff reported to a director (a veterinary scientist) responsible for all research activities in the ministry. The DoF was left without a substantive director at the ministerial level. It therefore became difficult to ensure that issues pertaining to fisheries received attention at a ministerial level dominated by functionaries without any knowledge of fisheries.

The above events were also to affect the legality of the entire co-management plan. In 1994, the DoF made proposals to its parent ministry to have the Fisheries Act amended to acknowledge the involvement of fishermen in management, and to make the ZMC and the IVMC legal entities. By 2000, however, this act had not yet been ratified by parliament and the department is currently unable to use the provisions of the Fisheries Act to compel other institutions, such as the local authority, to remit any levies to the ZMC and the IVMC as was agreed at the time of the plan's inception.

From 1998 onwards, the funds available for the co-management programme declined. In a review of its activities in the fishery, the Project resolved that its efforts would now focus on the joint management of the *kapenta* sector and would no longer financially support the activities in the artisanal sector.

Final outcomes

As not all the local actors felt that they had obtained what they had wanted from the new management plan, their involvement began to change. In December 1995, the *Kapenta* Fishermen's Association (KFA) informed the Zonal Management Committee (ZMC) that it was no longer in a position to continue supporting the zone as it was running low on funds. By the end of 1997, most *kapenta* operators were not as directly involved in the management process as they had been at the plan's inception. According to the KFA chairman, his association was not satisfied with the roles other institutions were playing in the new management plan. First, by allowing the artisanal fishermen to return to the islands, *kapenta* theft continued, and possibly even increased. The KFA blames individuals and institutions, such as the Member of Parliament and DoF, for allowing artisanal fishermen to go back to the islands.

In addition, the KFA is not satisfied with the type of service it receives from the local authority. The local authority has not provided any tangible services, such as the repair of roads leading to *kapenta* bases, despite the high fishing levies paid by the KFA to the local authority. Other institutions in the managerial arrangement, like the local authority and the DoF, have not remitted any funds to the ZMC as was agreed upon when the management plan was introduced. Consequently, the KFA sees no reason why it should be the only one supporting the management arrangement financially.

There were additional, little articulated, reasons why the KFA wanted to stop its support for the management plan. One of the main reasons was that some members of the KFA wanted to diversify their investments by having a stake in the tourist industry through the lease of islands. While some members of the KFA made investments on the islands and currently benefit from these, others were unable to make similar investments. While those who benefit from their investments continue to attend KFA meetings, those who have not benefited from the plan have reduced their involvement in it. This has lead to divisions within the KFA with some supporting the management plan while others have withdrawn altogether. Interviews with the chairman of the KFA indicated that his members would still be able to fully participate in the activities of the ZMC if the other institutions met their side of the bargain, namely the contribution of revenue to the ZMC.

Amongst those who were supposed to remit revenue to the ZMCs was the local authority. In 1996, a meeting was held to pass by-laws to legitimise the designated fishing villages, and authorise the local authority to remit funds to the ZMC. During the meeting, some local authority councillors expressed worries that remitting money to the ZMC would negatively affect the local authority's revenue base. According to one informant who attended the meeting, the local authority was unable to agree on making by-laws recognising the designated fishing villages because this would have automatically meant surrendering a large percentage of collected fish levies, an important source of revenue for the local authority to the ZMC. Instead, the local authority resolved to pass a by-law authorising it to collect fish levies without committing itself to surrendering money to the ZMC. There was not, in addition, any obligation for the new settlements to be recognised under the Local Authority Act.

The local authority also authorised the lease of the some of Kariba's islands, without informing other members of the ZMCs, such as the Department of Fisheries. When this factor was brought to the attention of the local authority, it responded that it did not have to consult other members of the ZMC because the local authority and the chief had jurisdiction over the islands (Joint Zonal Management Committee, 1995). The local authority was also against allowing fishermen to go back to the islands on the grounds that 'the re-arrangement of fishing grounds and free access by fishermen on any island were against the spirit of regrouping as it would re-introduce unlimited mobility by fishermen' (Joint Zonal Management Committee, 1995:11)

When local authority officials realised that the potential existed to obtain further revenue by allowing artisanal fishermen to return to the islands, however, a number of resolutions were passed. They explained that these were intended to control the conflict between the artisanal fishermen and the KFA over the use of islands (Sinazongwe District Council, 1995). The local authority agreed to monitor the movement of fishermen on to the islands by charging each fisherman ZK200[20] for fishing permits allowing them to stay overnight on the islands. Furthermore, the local authority also made the provision that only 25% of fishermen in a particular, designated, fishing village would be issued with island fishing permits at any given time. This arrangement was going to be monitored by the local authority with assistance from the KFA.

It was soon realised, however, that the local authority could only fulfil its objectives if it deployed levy collectors in the fishing camps to issue the overnight permits. The collectors were not employed and, consequently, the local authority and the ZMC were unable to stop or control the return of some artisanal fishermen to the islands.

Despite these problems, the local authority has still been able to collect some levies on fish leaving the fishery. None of this, however, is remitted to the ZMCs or the IVMCs as was envisaged in the new management plan because, the local authority says, the Fisheries Act has not yet been amended to allow it to do this. According to one local authority source, the money raised from the fish levy is used to meet some of the local authority's expenses in the face of reduced income from the government. At a meeting held to discuss the problems in the fishery, the local authority described the DoF as 'a big let down and not doing its job to have the Fisheries Act amended.'[21]

The results of the plan from the artisanal fishermen's perspective is mixed. By 1999, out of the seven fishing villages in Zone Two, three IVMCs were not functional and some of the fishermen had abandoned the designated fishing villages and gone back to their old settlements, while others had moved back to the islands. Some of these fishermen interviewed cited reasons such as distance between the fishing villages and the lakeshore as the reason for leaving the designated villages. Others complained of harassment from the local Tonga people as a reason for going back to the islands.

Other IVMCs, however, are functioning according to the plan. These IVMCs have been registered as welfare societies and this has enabled them to raise funds from new entrants into the fishery as well as through the implementation of the 'Certificate of Origin' certificates.[22] Such IVMCs still

---

[20] At current rates (2000) US$1 is equivalent to ZK3, 400.00

[21] Personal Communication with the Sinazongwe Local Authority Secretary, 21/10/1999.

[22] Between November and March every year all fisheries in the country except Lake Kariba are closed to fishing to allow for breeding. During this period the DoF issues a 'Certificate of Origin' to fish traders to certify that the fish in their possession is from Lake Kariba and not from other fisheries.

attract funding from members of the KFA who feel that their interests are best protected by participating in the activities of the ZMCs and the IVMCs. These IVMCs are also able to monitor and enforce fishing regulations in their fishing grounds.

Conclusion

The justification for the introduction of the new management arrangement was centred on a question of rights for fishermen, based on a concern for the lack of property rights. In turn, it was reasoned, this had caused there to be an increase in the number of fishermen as well as movement of fishermen from areas of low productivity to those where catches were high. This was said to be leading to a 'Tragedy of the Commons' scenario. Evidence of a tragedy was the decline in fish sizes and the increased use of illegal fishing methods.

One of the plan's outcomes was that local actors formed alliances with one another to use the plan to protect or advance their interests within the fishery. The new plan led to a redefinition of parts of the fishery around tourism interests, particularly on Lake Kariba's islands. In addition, the plan also exposed conflicts between local Tonga people and immigrant fishermen both within the fishery and over access to other equally important resources, such as land.

Most studies of fisheries management have tended to show that equity and efficiency can be obtained if fishermen are involved in the management of a fishery. The focus of such plans is on fishermen alone, and assumes that fishermen have similar interests and objectives within the fishery. This may be true in certain fisheries but, as this paper has shown, Lake Kariba is an arena with many actors with diverse interests. The local Tonga are part-time fishermen and part-time farmers, and decisions made to change the management of fishing affects their farming activities. What is important in fisheries management, especially in an inland fishery such as this one, is to understand the historical and socio-economic circumstances in which it is embedded. This may be useful in devising a management plan that encompasses competing claims to the resource. The homogeneity of expectations and the equity and efficiency, which are so much part of common property theories, do not seem to be borne out by the Lake Kariba experience. Local actors are dynamic and are likely to seize upon opportunities presented by the introduction of co-management to advance their interests at the expense of less organised and marginalised groups.

7

A Future Fraught: Precautionary, Participatory and Regional Outlooks for the Fisheries of Lake Tanganyika

John-Eric Reynolds,[23] *Hannu Mölsä,*[24] *and Ossi V. Lindqvist*

In this article we review the evolution and current status of the commercial fisheries of Lake Tanganyika, and ongoing attempts to develop a regional framework for their joint management by the littoral States of Burundi, the Democratic Republic of Congo (DRC), Tanzania, and Zambia. Preliminary sections of the article set the stage by placing the lake in its geographical, hydrobiological, and socio-economic context, outlining two major fisheries- and conservation-related projects that have recently been implemented – the Lake Tanganyika Research Project (LTR) and the Lake Tanganyika Biodiversity Project (LTBP) and summarising basic characteristics of the lake's fisheries and fishing communities. Principal management challenges within the Tanganyika fisheries are identified in a following section, with emphasis on hydrobiological, human welfare, and institutional and legal issues. A means for addressing these challenges through the Lake Tanganyika Regional Framework Fisheries Management Plan is then described. The Plan, based on the FAO Code of Conduct for Responsible Fisheries (CCRF), is a collaborative effort and involves the Lake Tanganyika Research Project, local communities, and national fisheries authorities. Its proposed implementation through the Tanganyika Regional Fisheries Programme is next reviewed. The programme would involve a series of initiatives to develop community management partnerships and local infrastructure and services within a harmonised regional framework of legislation and regulatory measures, improved monitoring capabilities, and institutional arrangements. The pursuit of these initiatives is anything but a straightforward process, because intended positive outcomes can only be achieved in the face of considerable difficulties and risks, including the possibility of exacerbating some of the very problems they are supposed to help resolve. As we argue in the final section, a 'Do Nothing' response to the Lake Tanganyika fisheries situation carries no such ambiguity. Its outcome, however, would be the most risky of all.

[23] Co-author for correspondence.

[24] Co-author for correspondence.

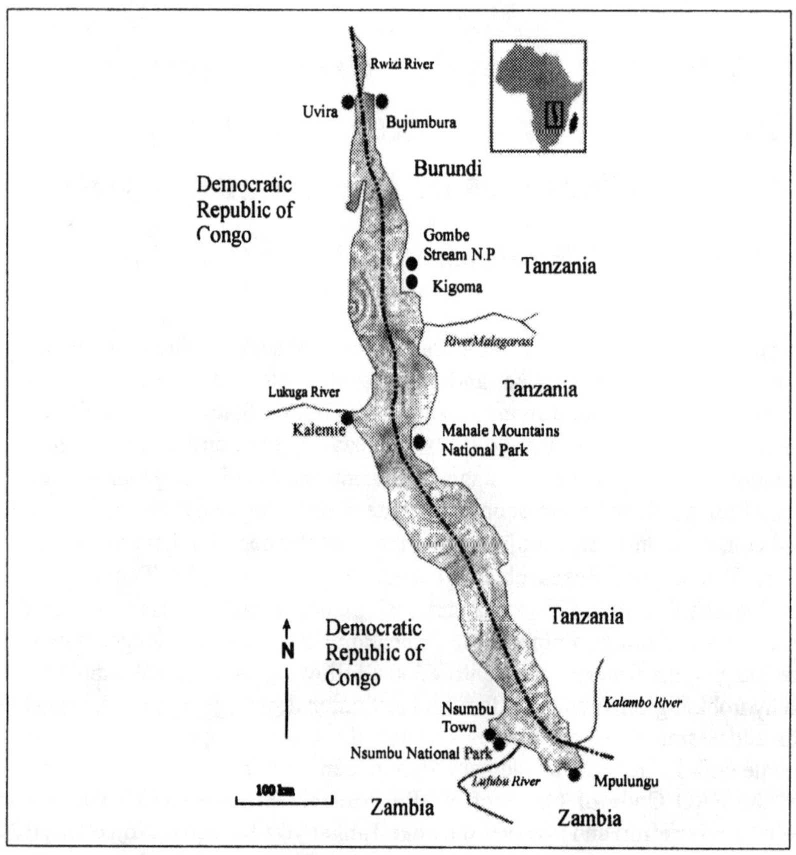

Figure 16: Lake Tanganyika region.
Source: Adapted with permission from Lindley (2000).

Tanganyika: the setting and the stakes

Lake Tanganyika is one of the most spectacular features of the African continent. Stretching 673 km in length, with an area of 32,900 km², a maximum depth of 1,470 m, and a volume of 18,880 km³, it is remarkable on many counts.[25] Entrenched within the Western Rift Valley between the countries of

---

[25] On the basis of these figures, Lake Tanganyika qualifies as: a) the longest lake in the world; b) the largest of Africa's Great Rift Valley lakes, the second largest (after Victoria) of all African lakes, and the fifth largest of the world's lakes; c) the deepest of all African lakes and the second deepest lake (after Baikal) in the world; and d) by cubic size, the greatest single reservoir of fresh water on the continent and the second greatest (after Baikal) in the world.

Burundi, the Democratic Republic of Congo (DRC), Tanzania, and Zambia (Figure 16), the lake's surface lies at an altitude of 773 m along a generally north-south axis between the narrow confines of the steep eastern and western escarpments of the Rift from 03°20'30"S to 08°48'30"S latitude. The lake averages almost 50 km in width and has a mean depth of 570 m (*vide* Coulter, 1991). Table 5 provides data on the allocation of surface area and shoreline frontage between each of the lacustrine states.

A further outstanding feature of Tanganyika, and one that it shares with the other Great Lakes of Malawi and Victoria, is a very high rate of endemism amongst its cichlid and other fishes as well as invertebrate populations. As in the case of the other lakes, this arises from its long geological history as an isolated basin (Lowe-McConnell 1969; Beadle 1981; Coulter, 1991, 1994).

Country	**Latitude Range**	**Lake Area** (km²)	**Lake Area** (%)	**Shoreline** (km)	**Shoreline** (%)
Burundi	03°20'30"S - 04°26'40"S	2,600 km²	8%	159 km	9%
DRC	03°21'00"S - 08°13'40"S	14,800 km²	45%	795 km	43%
Tanzania	04°26'00"S - 08°36'00"S	13,500 km²	41%	669 km	36%
Zambia	08°13'40"S - 08°48'30"S (West shore)	2,000 km²	6%	215 km	13%
Totals	03°20'30"S - 08°48'30"S	32,900 km²	100%	1,850 km	100%

Table 5: Lake Tanganyika: Division of national waters and shorelines

From a hydrobiological point of view, therefore, the inherent conservation value of Lake Tanganyika is enormous. Its qualities as a prime natural heritage asset stand out in other ways too. Lying between high escarpments, with extensive stretches of unspoilt beaches and rocky promontories, numerous bays, estuaries, and inshore islands, the near pristine waters of the lake border areas of wetlands, forest, and savannah containing a rich assemblage of tropical flora, terrestrial fauna, and bird life. All of these features contribute substantially to the lake's ecotourism potential.

From a socio-economic point of view, the lake's value is also enormous, since it provides a critical resource base for the human population of its basin and hinterlands. Tanganyika has from time immemorial provided a rich source of subsistence for lakeside communities in the form of fish and other aquatic life. The contemporary inhabitants of the basin area, estimated at some 10 million in number, depend heavily on the lake not only to yield food, but also to provide drinking water, income, and a medium for transport and commerce linking a vast area of eastern, central, and southern Africa. Many more millions of people residing within the wider trading orbit of the lake benefit from its resources indirectly, as occasional consumers of its fishery products and as users of the wide variety of other commodities produced within or conveyed through its basin (Hanek, 1994; Hanek and Craig, 1996).

There are obviously very compelling reasons for protecting the integrity of the Tanganyika lacustrine ecosystem and ensuring the sustainability of the fisheries it supports. The stakes are prodigious. As pressures on the lake and its resources continue to mount – driven largely by ever-expanding human populations and attendant settlement, cultivation, deforestation, urban/industrial water extraction and waste disposal, commercial development, and fishing activities – the risks of their loss become ever more apparent.

Lake Tanganyika Research Project

Concerns over such risks have been increasingly expressed over the last few decades. One major practical response to the situation has been the Food and Agriculture Organization (FAO)-executed Lake Tanganyika Research Project (LTR), which became fully operational in early 1992.[26] The project was established with the overall aim of promoting optimal fisheries management approaches to meet both human welfare and biological conservation imperatives, with particular attention to the commercial pelagic fishery – by far the most significant economic activity on and around the lake.

LTR's work programme initially focused on basic research related to hydrodynamics, limnology, fish and zooplankton biology, remote sensing, fish genetics, and fisheries statistics (Sarvala *et al.* 1999). Extensive studies on lake hydrodynamics, limnology and fish biology as well as catch surveys in LTR have showed the major role of climatic and physical factors in controlling the productivity and dynamics of biological communities (Huttula, 1997; Salonen *et al.*, 1999). The possible vulnerability of pelagic stocks to

[26] The full title of the LTR Project is 'Research for the Management of the Fisheries on Lake Tanganyika' (GCP/RAF/271/FIN). The Government of Finland has provided the bulk of project funding. Support was also received from the FAO Fish Code Programme (Interregional Programme for the Implementation of the Code of Conduct for Responsible Fisheries) under Project GCP/INT/648/NOR ('Fisheries Management') in order to complete fisheries planning activities initiated in the second phase of the LTR project. Limnological and fisheries monitoring work continues on a reduced scale under the responsibility of national personnel affiliated with LTR counterpart institutions within each country, including the Département des Eaux, Pêches et Pisciculture (DEPP) in Bujumbura (Burundi), the Centre de Rescherches en Hydrologie (CRH) in Uvira (DRC), the Tanzania Fisheries Research Institute (TAFIRI) station in Kigoma (Tanzania), and the Department of Fisheries station at Mpulungu (Zambia) (Mölsä *et al.*, 2001).

excess fishing capacity was also demonstrated (Mannini, 1998; Mölsä *et al.*, 1999; Sarvala *et al.*, 1999). A second phase of investigations concentrated on legal-institutional and socio-economic issues (Cacaud, 1996, 1999a, 1999b; Maembe, 1996; Reynolds and Hanek, 1997), and involved the formation of the LTR Fisheries Management Working Group. A lake-wide survey exercise mounted in 1997 collected a wide array of information on landing sites, fishers, and post-harvest operators in addition to attitudes and outlooks of local stakeholders in regard to fisheries problems and management measures. The second phase concluded in late 1998 with a 'community referenda' exercise that involved meetings between fisheries officials and local fisherfolk around the lake in order to review LTR research findings and obtain feedback on new regional management proposals (Reynolds, 1999a). A detailed work plan for the continuation of LTR's monitoring activities under national responsibility, as the Lake Tanganyika Fisheries Monitoring Programme (LTFMP), was also prepared at this time (Mannini, 1999). Complementary series of monitoring samples have now been collected for the study of inter-annual and seasonal patterns of lake limnology and fish recruitment (Mölsä *et al.*, 2001).

Following completion of the regional Framework Fisheries Management Plan (FFMP – Reynolds, 1999b), and its adoption by the four lacustrine states through the CIFA Committee for Lake Tanganyika, the African Development Bank (AfDB) and FAO were jointly requested to elaborate an implementation programme. This resulted in the preparation of a feasibility study for the Lake Tanganyika Regional Fisheries Programme (TREFIP) and a follow-up environmental impact assessment (Magnet *et al.*, 2000; Reynolds and Mölsä, 2000).

Lake Tanganyika Biodiversity Project

From 1995 to mid-2000 another major undertaking, the Lake Tanganyika Biodiversity Project (LTBP) operated along lines that were in principle largely complementary to LTR's pelagic fisheries-related investigations. LTBP's remit was to address wider, basin-scale problems of pollution control, conservation, and the maintenance of biodiversity, with a view towards establishing a regional long-term programme of management interventions (LTBP 2000a, 2000b, 2000c; West, 2001). Special study teams worked in the areas of biodiversity, pollution, sedimentation, fishing practices, and socio-economics to investigate the state of the lacustrine and basin environment and the factors affecting its sustainability. Principal LTBP outputs have been a series of technical reports in the special study areas and, on this basis, the preparation of a Strategic Action Programme (SAP) for the Sustainable Management of Lake Tanganyika and a draft 'Convention on the Sustainable Management of Lake Tanganyika.' The draft Convention now awaits ratification by the four lacustrine states (the Convention comes into effect when two of the states have ratified it). It is not at the moment clear how long this process will take, but it is likely to be extend over many years (UNDP/GEF 2001:5).

Efforts are now (late-2001) underway to develop regional and national project proposals to implement the SAP and for this purpose a one-year planning phase for the initial set of projects is being executed. SAP implementation projects will supposedly address specific 'hot spots' and sources of transboundary problems, especially in the area of pollution control, habitat destruction, sedimentation, and over-fishing. GEF/UNDP, AfDB, and FAO have agreed in principle that close collaboration is called for and efforts should be made to integrate the SAP and TREFIP proposals into a common programme of action.

Basic fisheries characteristics

Fishing in the lake expanded enormously in the course of the 20th century, especially since the 1950s. Its basic character as a subsistence-oriented activity prosecuted with relatively simple tools of largely local manufacture was transformed to a commercially-oriented one requiring substantial inputs of gear and equipment imported from industrial countries overseas. This transformation was associated with the expansion of human population and settlement within the lake basin, and the growth of marketing outlets including lakeshore commercial centres and, especially, the Copper Belt complex of large towns in northern Zambia and the mining areas of south-eastern DRC. Its technological basis was provided by the introduction of various innovations, such as paraffin oil (kerosene) pressure lamps for night-fishing, synthetic netting material, motorised fishing craft, and chilling and freezing facilities for fish processing.

Fishing of inshore demersal stocks (both cichlid and non-cichlid species) is widespread but, at least insofar as official catch returns and other records indicate, of comparatively minor importance in terms of volume and commercial value (cf. Coulter, 1991; Pearce 1985a, 1985b, 1985c), accounting for less than two percent of total annual catches from the lake (Coenen and Nikometze, 1994; Coenen, 1995). It is likely, however, that a substantial portion of inshore fishing activities goes unreported in official statistical returns. Historically, the commercial fisheries have mainly exploited two small pelagic species, the schooling clupeid 'sardines,'[27] *Limnothrissa miodon* and *Stolothrissa tanganicae*, together with their major predators, all centropomids of the genus *Lates*. Most pelagic fishing is based on light attraction at night,

[27] Known variously as '*ndagala*' (Burundi and DRC), '*dagaa*' (Tanzania), or '*kapenta*' (Zambia).

and is thus dependent on weather and lunar conditions. Little fishing takes place during full moon periods, when artificial light sources are not very effective. Windy nights also discourage activity, as the surface of the water can become very rough and fish schools stay at deeper levels (Coulter, 1991; Coenen *et al.* 1998).

Fisheries sub-sectors

The Tanganyika fisheries are generally classified into three basic types, according to gear kits, equipment, and scale of operation involved. These comprise the 'traditional,' 'artisanal,' and 'industrial' sub-sectors.[28] The 'traditional' fishery is prosecuted in waters close to the shore. It has, historically, been based on the use of *lusenga* or scoop nets and light attraction (formerly with flares made of bundled cane, now with lamps) for the harvest of clupeids, and gill nets, long lines, hand lines, traps, spears, and poisons for the capture of demersal species.[29] Fishing units comprise one or two persons operating with paddle-propelled dugout canoes or with plank-built canoes propelled by paddle and (in some places) lateen-rigged sails. The use of *lusenga* nets has drastically declined in recent decades owing to widespread adoption of more efficient lift net and beach seine gear. Most traditional fishing nowadays is prosecuted with small-mesh gill nets. Another method that is becoming increasingly popular around the Kigoma Region is line jigging for *L. stappersii*. Jigging is carried out during daylight hours in deeper waters usually within 5 km of the shoreline, in units comprising one to three persons. Canoes sail out to the fishing areas on the morning offshore wind, and return in the late afternoon with the onshore wind. Recent studies have also demonstrated that, whilst remaining very much a 'sideshow' of the commercial pelagic fishery, the inshore traditional fishery has nevertheless been vastly under-appreciated in both its size and potentially negative effects (Lindley, 2000). There are also indications that traditional fishing generally is on the increase in recent years, because of growing economic decline and instability across the region. As conditions of food and civil security become more difficult, more people are attracted to fishing for subsistence as well as a potential source of income. Traditional gear and equipment is usually cheaper to purchase and operate, and easier to repair or replace, than is the case with artisanal fishing tools (Lindley, 2000; West, 2001).

[28] The 'traditional' – 'artisanal' distinction seems somewhat arbitrary nowadays. Historically 'traditional' fishing was primarily for subsistence purposes and employed relatively simple and often homemade gear. Nowadays, traditional operators often sell their catch if the opportunity arises, and there is heavier reliance on purchased inputs.

[29] Lindley (2000) reports the existence of more than 50 different types of fishing gear in use around the lakeshore, though only a few of these (purse seines, beach seines, lift nets, gill nets, and jigged lines) are significant in terms of distribution and fish harvests.

'Artisanal' fishing is primarily carried out for commercial purposes using lift nets, '*chiromila*' seines (ring nets), and beach seines. The artisanal fishery has grown immensely from the late 1950s, when the technique of lift netting from catamaran rigs was first introduced in the northern portion of the lake. Lift net units are equipped with 4 to 8 pressure lamps and operated by a four to six person team. The level of motorisation (outboard engines) for the overall fleet of small craft on the lake remains low, at something less than 10%. The use of engines is restricted almost entirely to artisanal units.

'Industrial' fishing units each comprise a large (16-20m) diesel-powered steel main vessel, a smaller net-setting vessel, and three or more light boats, each requiring a crew of 20 to 40 persons to operate. The industrial fishery traces back to the mid-1950s, when Greek nationals introduced purse seining in Burundian waters. Purse seine units operated from larger ports throughout the lake in subsequent years, but are now concentrated in the southern portion. Of the 13 industrial units active in Burundi in 1992, only two were enumerated as active in the 1995 LTR Frame Survey (FS) and only one was known to be active as of mid-2000. The remainder have either been decommissioned or have been shifted to Zambia in the south of the lake. The DRC has witnessed a similar decline in purse seining operations based in Kalemie and Moba, though this probably owes as much to political instability as to adverse fishing conditions. In Tanzania, the industrial fishery never developed to the same extent as elsewhere, but here too purse seining has fallen off in recent years. Of the four operational units enumerated in 1995, none are active at the present time.

Production and trade

FS data collected in 1992 (Coenen, 1995) and 1995 (Paffen *et al.*, 1997) indicate that there are some 45,000 fishers and a fleet of over 18,000 small craft active in the traditional and artisanal sectors of Lake Tanganyika. Together, these sectors are thought to account for over 90% of annual fish harvests, which in recent years have been estimated to vary in the range of 165,000 - 200,000 t. Harvests comprise principally the two clupeids (ca. 65% by weight) and *L. stappersi* (ca. 30% by weight), and it is estimated that they are distributed between the littoral countries roughly in the order, if not exact proportion, of each country's share of the total lake area.

Due to the remote location of much of the Lake Tanganyika coastline, most of the catch landed by artisanal and traditional operators is processed by simple sun-drying or smoking before being shipped to markets. Around many larger landing sites, increased demand for fuel wood to supply fish smoking operations has contributed to extensive loss of forest cover on hillsides above the lake, leading to sheet and gully erosion and siltation of influent streams and near shore waters.[30] The industrial fishing companies in Mpulungu (Zambia) are equipped with far more capital- and energy-intensive facilities and market their catches, along with some fish purchased from local artisanal operators, mostly as frozen product. Although the precise extent of the Tanganyika fish trade is difficult to document, it reaches areas well beyond the lake. Long-established market outlets include the mining districts of south-eastern DRC and the Zambian copper belt. In Tanzania, the railway link from Kigoma to Dar es Salaam serves as a distribution corridor between the lake and the Indian Ocean coast. A steady supply of Tanganyika fish products also flows north and west, through the DRC and Rwanda and sometimes as far afield as the Central African Republic. Lately, substantial quantities of dried fish have reportedly been distributed by relief agencies as emergency food assistance for refugee camps and repatriation schemes (Reynolds and Hanek, 1997; Reynolds, 1998).

Such volumes of harvest and trade represent commercial activity that can be valued in the order of tens of millions of US dollars, and rank Lake Tanganyika as one of the largest inland fisheries on the continent (FAO, 1995a).[31] The lake's role as food provider has become all the more critical for the general East-Central Africa region in recent decades owing to steadily increasing human populations and the disruption of crop and livestock production brought on by episodes of civil turmoil, economic displacement, drought, and the incidence of HIV/AIDS.

Local stakeholders and communities

There are nearly 800 fishing communities of various sizes distributed around the coastline of Lake Tanganyika. Members of these local communities represent the primary bloc of Tanganyika resource stakeholders, as the lake is their immediate source of livelihood and sustenance. Those most directly involved in the fisheries include harvest sector workers and owners of productive equipment (gear, canoes, etc.), post-harvest sector fish processors and traders, and providers of various support services (craft repair, spares, fuel, food stands, lodging, etc.). A far greater number of coastal people rely on the lake's resources as consumers for whom fish is a crucial source of

---

[30] Evidence of increased human pressure on basin resources is shown dramatically in terms of deforestation, as land is cleared to provide fuel wood, charcoal, and cultivation, grazing, and settlement areas. Analysis of Landsat images indicates that 40-60% of original forest cover in the central basin has been cleared. In the northern basin, the figure is almost 100% (Cohen, 1991, cited in West, 2001).

[31] It is ranked second to Lake Victoria in both volume and value terms. The recent AfDB/FAO mission estimated that annual earnings from Tanganyika fisheries in recent years range from US$ 80 to 100 million (Magnet *et al.*, 2000).

animal protein food, and as family and household dependants of fisherfolk. Taking all these categories into account, it can be estimated that the welfare of some one million lake dwellers – perhaps one tenth of the entire Tanganyika basin population – is more or less directly tied to the fate of the fisheries. Some key general socio-economic characteristics of local artisanal and traditional fishers and of processors and traders are summarised at a very general (lake-wide) level in Table 6 (see Meadows and Zwick, 2000; Reynolds, 1997a, 1997b, 1997c, 1997d; Reynolds and Hanek, 1997 for more detailed information).

The 1997 LTR sample survey data of lakeshore communities revealed an overall picture of poor physical infrastructure and a dearth of basic social services and amenities (Reynolds and Hanek, 1997). They also indicate that fishers in these communities live in relatively straitened conditions. For the most part, they command very limited resources in terms of formal education (completion of primary school), savings, earning power, and ownership of land, housing, and consumer goods, even as they strive to provide for family and household dependants.

Characteristic	Fishers (Unit owners & crew members	Fish Processors &
Gender	Almost all male (very few women boat owners reported)	Mixed. Women constitute majority of processor/trader population in Zambia and parts of DRC.
Age range	Mostly 18-50 years. Owners generally older (majority >30 years) than crew (majority <30 years).	Mostly 18-40 years – younger as group than fishers.
Formal education	Generally have not completed primary level.	Generally have not completed primary level, with women registering lower than men.
Residential history	Mostly native-born to current landing site bases, but high incidence of immigrants.	Mostly immigrant.
Secondary/ the r employment	Subsistence or combined food crop/cash crop farming.	Fishing related (usually gear owner) and farming.

Table 6: Basic characteristics of Lake Tanganyika fisherfolk. **Source:** 1997 LTR Lake-wide Community Survey data of fishers (n = 923) and post-harvest operators (n = 431) at 66 landing sites (geographically stratified by country; see Reynolds and Paffen 1997).

Lake Tanganyika's fishing communities thus share in the wider regional conditions which, measured by standard 'quality of life' indices, define East-Central African countries as amongst the world's most poverty-stricken and underdeveloped (World Bank, 1999; cf. West 2001). Nevertheless, fisheries remain a fairly attractive mode of employment for Tanganyika basin inhabitants. Income prospects may not be spectacular, but they compare favourably to those that can be garnered in other sectors of the economy, especially within rural areas.[32] This set of circumstances not only serves to keep people involved in fisheries work, but acts to draw in others as well.

Major management challenges

LTR hydrobiological, socio-economic and legal-institutional investigations over the 1992-1998 period emphatically confirm the view that the Tanganyika fisheries are headed for trouble – i.e. deterioration towards non-sustainability. Briefly, this multidisciplinary set of 'readings' on the fisheries situation is as follows.

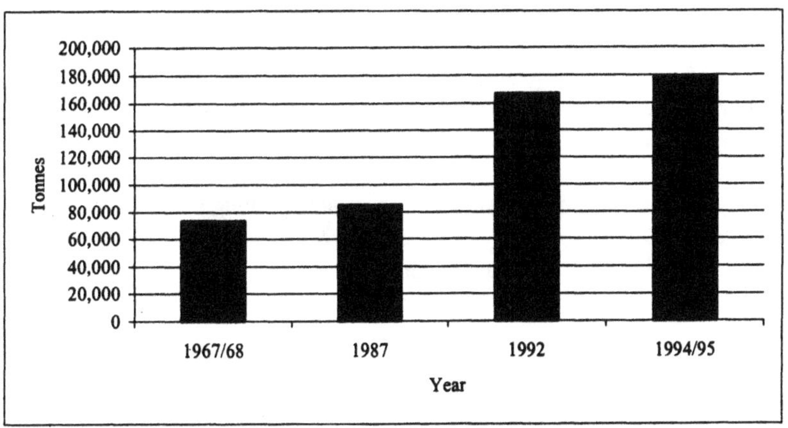

Figure 17: Nominal catch (t), L. Tanganyika, 1967-1995.

[32] Estimates based on World Bank (1999). It should be noted that even the highest Tanganyika fisheries-related incomes rarely fall beyond the range of US$500 – 1,000 (equivalent) a year. Average annual incomes amount to the equivalent of a few hundred US dollars.

Hydrobiological issues: changes in stocks and yields

Reported annual catches on Lake Tanganyika have shown a marked upward trend in the period from the mid-1960s to the mid-1990s (Figure 17).

Estimates based on LTR catch-per-unit-effort (CPUE) calculations (Coenen *et al.*, 1998) yield a lake-wide production figure of 196,570 t for the mid-1990s – rather higher than the reported figure of 178,700 t). This upward trend occurs in a context of changes in gear technology, declining CPUE, and other fundamental developments that are summarised in Table 7. Some observers would no doubt question this inventory of developments on the grounds that it highlights negative effects of fisheries exploitation and is influenced by 'Malthusian' tendencies in its concern with human population loading of the Tanganyika basin in association with various forms of environmental degradation.

With respect to the first issue, commentators have recently stressed the importance of abiotic or climatic factors in driving fish stock fluctuations and the changing fortunes of the fisheries on various African inland water bodies, and have pointed out the inappropriateness of management regimes that seek to limit access or in other respects regulate fishing mortality (e.g. Sarch and Allison, 2000; Jul-Larsen, 2000).[33]

In our estimation such arguments, whilst correctly noting that large-scale climatic events and cycles have profound and far-reaching effects on fisheries fortunes, effects that have generally been underestimated or unrecognised in the past, nevertheless tend to err too much in the other direction by undervaluing anthropogenic causes of fish stock fluctuations and declines[34].

As far as Lake Tanganyika is concerned, whilst abiotic effects on ecosystem productivity must certainly be taken into account as playing a very substantial role (Plisnier, 1997; Coenen *et al.*, 1998), the role of artisanal and industrial fishers and their various gear in influencing the composition of fish stocks and catches, has also been clearly demonstrated.

---

[33] For reviews of global climate changes on climate variability and consequent hydrophysical events in Southern and Eastern Africa see e.g. Cole (2000) and van Jaarsveld and Chown (2001).

[34] This includes the under-valuation of possible long-term ecosystem disruption as stocks of higher value, longer-lived species dwindle and stocks of previously non- or lightly exploited low value forage fish become the targets of effort. Such arguments may also overlook or underestimate the anthropogenic forces that induce changes across entire natural resource systems and thereby threaten broader ecosystem integrity, as for example could occur where protracted and unrestrained opportunistic exploitation of wetland and forest resources (clearance or conversion for settlement and cultivation) results in loss of nursery areas and fish refugia and increased sedimentation of littoral waters.

Fisheries-related developments	**Fisheries-related effects**
Shift from traditional methods to artisanal lift netting and beach	Overall average single unit efficiency increase from 3 to 14 t. yr^{-1} (Coenen 1995).
Selective purse seining pressure on stocks of *L. mariae, L. microlepis,* and *L. angustifrons* (Coulter, 1970, 1991).	Stock reduction; simplified composition of commercial catch (two clupeids and *L. stappersii*) found today.
Retirement/withdrawal of industrial units from northern waters (competition from artisanal lift nets, civil strife, other factors). Migration to southern waters.	Zambian sector: 3 active units 15 years ago versus 30 now, almost exclusively targeting *L. stappersii* (Coenen *et al.*, 1998; Mannini 1998).
Drop in CPUE (kg/boat/night), best documented for industrial operations during the early to mid-1990s (Coenen *et al.*, 1998).	Burundi: from 166 kg to 111 kg; Zambia: 877 kg to 535 kg; DRC: from 780-950 kg (early 1990s) to 433 kg.
Increased duration of fishing trips for southern purse seiners (Coenen *et al.*, 1998; Mannini, 1998).	With progressive CPUE declines, indicates decrease of catchable stock, possible over-exploitation of *L. stappersii* owing to uncontrolled growth of the industrial fishery.
High juvenile content in northern catches of *L. stappersii*, lower proportions of *L. stappersii* in overall catches.	Signals possible excess fishing pressure on *L. stappersii* in north (Mannini 1998).
High juvenile content in northern catches of *S. tanganicae*.	Signals possible over fishing of *S. tanganicae* stocks in north (Mannini 1998).
Extensive use of unselective beach seines.	Destructive to juvenile *L. miodon* (shallow, inshore nursery grounds); also to cichlid community (Mannini 1998).
General growth of smallcraft fleet from ca. 12,800 units in 1980s (Coulter, 1991) to ca. 18,000 units in 1990s (Paffen *et al.*, 1997). Increased effort directed at inshore demersal fishery, marked especially by extensive use of gillnets.	A collapse or serious decline of the inshore fishery would likely lead to transfer of effort to the pelagic fishery, and vice versa, with attendant complications (cf. Lindley 2000).
High human population growth rates (2.5 – 4.3%) within the lake basin and expansion of land clearance for settlement and cultivation (World Bank 1999; West 2001).	Population doubling times of between 17 to 30 years; increased demand for fish, and recruitment of newcomers to employment in the harvest and post-harvest sectors; environmental degradation and fish habitat effects.
Increased demand for fuelwood supplies for fish smoking operations (Reynolds and Mölsä 2000).	Environmental degradation; localised erosion and siltation of influent streams, nearshore waters.

Table 7: Major development and effects in Tanganyika fisheries, ca. 1950 to present. *Reynolds, Mölsä and Lindqvist*

With respect to the second issue, it is a well established argument that the relationships between population growth, poverty, and environmental degradation are variable and can be heavily mediated by circumstance. Allison (2000) reminds us that Boserup's work (1965) in particular has been influential in demonstrating that these factors should not be linked automatically together in some ineluctable sequence of cause and effect. Factors of land surplus, technology, and labour availability, amongst other things, may interact to favour extensive production strategies (e.g., land clearance and tree cutting for new cultivation and settlement) in some instances, and intensive ones (involving, e.g., build-up of soil fertility, terracing, erosion control, tree planting) in others.[35]

This having been said, we find it difficult to escape the conclusion that increasing human population loads around the Tanganyika littoral are having extremely serious consequences in that they are giving rise to unsustainable conditions of natural resource exploitation pressure and environmental risk. It is also difficult to accept as very reassuring the observation that the slopes above the shoreline of Lake Tanganyika have historically been subject to incursive, high angle cultivation, or that, in a somewhat contradictory vein, present cultivation practices on these same steep slopes may be linked to the prevailing climate of political instability within the region (Allison, 2000). What needs to be borne in mind through all of the debate relating to anthropogenic influences on ecosystem productivity and the links between population growth and environmental degradation is precisely the *unprecedented* nature of present events of change. The scope and scale of the current exploitation of fisheries and other resources within the Tanganyika basin, and the ever increasing size of the human population that depends upon them, render the contemporary lake very different from what it was heretofore. This is not only a simple temporal difference between 'then' and 'now.' It is a wholesale difference – of quantity, of quality, and even of place.

Human welfare issues

Recent observations related to the socio-economic and community welfare dimensions of Tanganyika fisheries identify the following major issue areas (Reynolds and Hanek, 1997; Reynolds, 1999a; Magnet *et al.*, 2000; cf. Meadows and Zwick, 2000; West, 2001):

---

[35] In the East African context, one of the mediating factors that should be taken into account when evaluating environmental conservation developments in densely settled rural areas (e.g. tree planting on private land, use of erosion control techniques) is the high incidence of labour out-migration, and the consequent availability of remittances and non-farm sources of income to support such developments. Whether such positive outcomes would ensue in the absence of these externally-generated resources is another question.

(a) *Fish supply and demand.* The contribution of fish to the total animal protein supply within the four littoral states currently ranges from some 25 to 40% (Gréboval *et al.*, 1994), but this crucial nutritional welfare role is becoming ever more difficult to sustain in the face of unremitting population growth across East-Central Africa (World Bank, 1999).

(b) *Resource access.* The current open access regime is not sustainable in the long run because it leads to a 'race to fish'. As pressure on the resource mounts, there are fewer returns available to share, and everybody then rushes to capture as much as they can before others do (FAO, 1997: 52).

(c) *Local empowerment.* Although existing legislation in some cases provides for consultation between administrators and local representatives of fisher interests (Cacaud, 1999a), and although fisher committees exist at various landing sites (Reynolds and Hanek, 1997), and claims of effective project partnership structures have been registered (West, 2001: 69), *de facto* local community participation in resource management decision-making and follow-up has been very minimal.

(d) *Equity.* Prospects for developing modalities for sustainable management and conservation are clouded by widespread socio-economic inequalities (Reynolds 1999a). These include: relations between fishing unit owners and fish workers, gender-based differences and relations between artisanal and traditional fishers on the one hand and the industrial fishing companies on the other (conflicts over fishing grounds, alleged over-fishing, alleged favouritism by political leaders).

(e) *Piracy.* Incidents of piracy on the lake have shown an alarming increase since the early 1990s, in association with political and economic chaos in the Congo, civil unrest in Burundi, population dislocations and refugee resettlement, and widespread availability of sophisticated small arms obtained from war zones (Magnet *et al.* 2000). The usual pattern involves armed gangs setting upon lift netters at night. Nets, equipment, catches, and personal belongings are hijacked at gunpoint, and taken off to be hidden across the lake or at some point further along the adjacent shore.

Institutional and legal issues

Evaluation of institutional and legal aspects of the Tanganyika fisheries is based on organisational and legislative framework studies (Hanek, 1994; Maembe, 1996; Cacaud, 1996, 1999b, 1999c), monitoring and statistical data collection in association with national research and administrative authorities (Coenen, 1994, 1995; Coenen *et al.*, 1998; Paffen *et al.*, 1997; Mannini, 1999), and socio-economic investigations (Reynolds and Hanek, 1997; Reynolds, 1999a). Mölsä *et al.* (1999) have summarised key points, and highlighted the following features and deficiencies:

(a) *Policy orientation.* The four lacustrine states share a stated policy orientation towards social welfare objectives, whilst recognising a requirement to secure sustainable resource use over the long term. At the same time, there is in all cases a basic lack of institutional means to achieve policy objectives.

(b) *Budget shortfalls.* Chronic under-funding seriously cripples the performance of national fisheries departments and research agencies, almost to the point of operational paralysis in some cases. Departments are unable to provide adequate extension or monitoring, control, and surveillance (MCS) services. National fisheries researchers are hard-pressed to fulfil their roles as management and conservation technical advisers, and rely to a great extent on outside sources of funding.

(c) *Harmonisation of legislative frameworks.* A consistent set of regulations needs to be developed for the management of stocks that are in reality unitary populations, not territorially divided sub-groups. Fisheries law derives from colonial era decrees in some cases, and there is a general need for update and overhaul.[36]

(d) *Regional management co-ordination.* The CIFA Committee for Lake Tanganyika serves as a forum for technical discussions between the four states on the management of the fisheries. It is constituted as a consultative rather than executive body, however, and meets (at most) only once every two years. Although all four states are involved in the Committee, none of them provides legislative authority for participation in a fully-fledged regional management authority.

(e) *Enforcement and compliance.* Regulations are widely ignored in practice, since they are either insufficiently enforced or not enforced at all. Co-management arrangements, through which local stakeholders would play an active role in regulatory decision-making and compliance assurance, have not been strong features of the 'top-down' institutional culture of national fisheries authorities.

The legal-institutional picture in the DRC is particularly dire, owing to years of political turmoil and economic malaise both before and after the overthrow of the Mobutu regime in Kinshasa by the Kabila forces in 1997. Subsequent conflicts that have engulfed large portions of the eastern half of the country have all but obliterated the institutional and legal expressions of state management functions for the Lake Tanganyika fisheries.[37] Civil unrest

[36] New draft fisheries legislation is under development in Zambia, Tanzania, and Burundi.

[37] Staff at the Centre de Recherche en Hydrobiologie (CRH) at Uvira, on the northern tip of the lake, have remained relatively active in fisheries research activities in recent years through support from LTR and LTBP.

has also disrupted fishing activity and severely impaired management functions in Burundi. The military authorities have closed down nearly all landing sites, concentrating activities within a few larger centres, and from time to time restrict fishing operations altogether. During most of 1996 a complete ban was imposed on night fishing. Travel along the shoreline is heavily policed.

Management initiatives for fisheries futures

'Whole ecosystem' approaches to management and the Code of Conduct for Responsible Fisheries

The Framework Fisheries Management Plan (FFMP) for Lake Tanganyika, developed through collaboration between the LTR Project Management Working Group and local fishing communities and fisheries authorities, is based on guidelines laid out in the FAO *Code of Conduct for Responsible Fisheries*, or CCRF (FAO, 1995b).[38] The Code reflects the broad paradigm shift in fisheries management thinking that has occurred over the last twenty-five years or so. As the extensive contemporary literature on the subject attests, this shift has been stimulated by widespread realisation of the inadequacies of 'conventional' management approaches in the face of ongoing fisheries crises and collapses across the world. Such inadequacies are commonly identified in three critical areas:

(a) The 'command and control' nature of decision-making and regulatory mechanisms means that management is something that is imposed upon rather than negotiated or developed with industry participants. Conformance to particular measures is thus likely to depend more on the application of official sanctions than on voluntary commitment and self-regulation.

(b) There is heavy dependence on simple 'stock assessment driven' biological models and other methods that, however sound in and of themselves, are inadequate monitoring and diagnostic tools for addressing the immensely complex, dynamic, and often unpredictable processes that underlie ecosystem functioning and fish production, including species interactions.

(c) Finally, there is a failure to take into sufficient account the pluralistic nature of fisheries systems, which involve not only physical and hydrobiological factors but socio-economic, cultural, and institutional aspects as well.

---

[38] The CCRF was adopted as a voluntary instrument by member states at the 28th FAO Conference in October 1995, having been under development since 1991 when the need for a set of international guidelines was first noted during the 19th Session of the FAO Committee on Fisheries.

'New paradigm' thinking links biological conservation as a necessary condition for fisheries sustainability with other conceptual and practical requirements for successful management. This 'ecosystem-based' orientation towards understanding and addressing problems of resource base and environmental preservation is a core principle of the CCRF. Besides its holistic outlook, a fundamental characteristic of ecosystem management thinking is that it is *processual.* Management is construed as inherently dynamic and adaptive in the face of complex and changing circumstances of potentiality and constraint. These provide the bio-physical and socio-economic reference points in terms of which the decision-making, MCS, consultation, review and reporting transactions of management should continually be carried out in order to achieve desired outcomes. Continued productivity of the resource, the most fundamental of such outcomes, can only be secured by controlling fishing mortality through one means or another – in other words, '...by regulating the amount of fish caught, when they are caught, and the size and age at which they are caught' (FAO 1997: 45).

Tools for the regulation of fishing mortality, which singly or in combinations have different efficiencies, impacts on fishers, and implications for MCS measures, are well known and in some cases have very long histories. Technical measures for the regulation of fishing mortality include restrictions of gear type, characteristics (e.g. net mesh size), and operation (e.g. 'active' gill netting). Other technical measures include area and time restrictions that define open and closed 'windows' for the application of fishing effort, as for example with 'no fishing zones' in known breeding and nursery grounds during particular months, or with aquatic reserves or protected areas (PAs) for the conservation of critical habitat and biomass. Input control can be used to regulate fishing mortality through the imposition of limits on fishing capacity and effort. Typical mechanisms include licensing ceilings, individual effort quotas on fishing units, and the use of technical specifications to limit the harvesting power of vessels and/or their gear kits.

The 'new paradigm' management approach embodied in the CCRF fully recognises the appropriateness of these standard regulatory tools in given circumstances, but proposes that their effectiveness will depend upon the application of safeguards to allow for ecosystem uncertainties and risks, and further upon the involvement of resource users and stakeholders in their planning and implementation. The Code thus highlights two additional key principles of fisheries sustainability. These are:

(a) adherence to the 'precautionary approach' in resource decision making; and
(b) the use of management partnerships wherein resource use and conservation decision making and stewardship responsibilities are shared between state fisheries authorities, local fishers, and other stakeholders.

In accordance with Article 15 of the Rio Declaration of the UN Conference on Environment and Development (UNCED), '...the precautionary approach to fisheries recognises that changes in fisheries systems are only slowly reversible, difficult to control, not well understood, and subject to changing environment and human values' (FAO, 1996a: 6). Basically it entails the use of conservative, risk averse, or 'do no harm' exploitation and development strategies in the face of system uncertainty.

CCRF *Technical Guidelines* characterise 'management in partnership' as embracing '...the various arrangements which formally recognize the sharing of fisheries management responsibility and accountability between a fisheries management authority and institutions either public, such as a local government authority, or private, such as a group of interested parties' (FAO 1997: 55).

The Tanganyika Regional Framework Fisheries Management Plan (FFMP)
The FFMP was developed as a basic set of recommended policy initiatives and practical actions intended to:

(a) help set the overall regional fisheries management stage (prerequisites for other actions); and
(b) address problems demanding immediate attention because of potentially serious sustainability impacts.

The framework proposes that initiatives be taken in a co-ordinated way at regional, national, and local levels as summarised in Table 8.

MANAGEMENT ISSUES	FFMP RECOMMENDED ACTION
A. Overall policy matrix	(a) Adoption of CCRF by competent authorities of respective States as the policy matrix for the shared fisheries of Lake Tanganyika, noting that this implies, *inter alia*, (b) the use of a 'whole ecosystem' approach to the lake and its resource, adherence to the 'precautionary principle,' and the use of partnership arrangements to secure management objectives; (c) reliance on adaptive or interactive management practices that permit adjustments in fishing pressure and the flexible application of treatments appropriate to different circumstances encountered around the lakeshore; (d) the establishment and operation of a multi-disciplinary monitoring capability for measurement of continuity and change across bio-physical and socio-economic dimensions; (e) Use of integrated strategies to accommodate complex interactions and possible conflicts between fishing and non-fishing activities, and, at national and regional 'macro-levels,' moves to foster economic diversification to reduce pressure on the fishery resource base.
B. Partnership and resource access	(a) Facilitate local stakeholder involvement in management structures and operational arrangements. (b) Provide for community outreach activities with a strong environmental education component. (c) Allocate control of access through community-based arrangements.
C. Institutional and legal modifications	(a) Establish a legal framework for local community involvement in fisheries management decision-making and enforcement activities. (b) Overhaul and update existing legislation and provide regulations specific to Lake Tanganyika, as appropriate. (c) Establish a permanent and effective lakewide/ regional management agency.
D. Possible immediate measures to regulate fishing	(a) Initiate gradual process leading to total retirement/phasing out of beach seining on the lake. (b) Encourage gradual reduction of effort in the purse seine fishery in the southern part of the lake to levels that prevailed ten years ago, either through unit retirement or transfer to other fishing zones. (c) Use closed or 'off-limits' reserve areas as a regulatory device. (d) Establish and enforce licensing ceilings, particularly in respect to industrial and liftnet units.

Table 8: Lake Tanganyika Framework Fisheries Management Plan: Summary elements

FFMP Implementation: Tanganyika Regional Fisheries Programme (TREFIP)
TREFIP was designed as a five-year undertaking to give effect to the FFMP through a set of four national project components operating under the co-ordination of a Regional Programme Implementation Unit. Major initiatives in six interrelated areas would include selected development activities aimed at building performance and value output from the fisheries without increasing pressure on existing resources, ameliorating environmental impacts of fisheries harvest and post-harvest operations and other natural resource-extraction activities linked with lakeshore settlements, improving infrastructure and social welfare amenities, and fostering economic diversification.

1) Promotion of management partnerships
This initiative will support:

(a) the establishment and operation of Community Fisheries Management Zones (CFMZs) and Local Fisheries Councils (LFCs) to be linked together through the Lake Tanganyika National Fisheries Councils (NFCs) and a Lake Tanganyika Regional Fisheries Council;
(b) the establishment and operation of a micro-credit scheme; and
(c) the introduction of appropriate gear and fleet restructuring.

The programme plan calls for a pilot CFMZ and LFC scheme that would eventually cover 200 fishing villages (ca. 25% of the lake-wide total). In addition to promoting the community education and outreach activities needed to build levels of environmental consciousness and receptivity to measures for the regulation of resource access and exploitation, the programme would facilitate direct partnership actions in support of responsible fishing practices and improvement of local welfare. These include, amongst other things, the formulation of appropriate measures to control access within community-based fishing zones, and of the compliance mechanisms needed to ensure the effectiveness of these measures.

New forms of licence and fish levy revenue allocation to both LFCs and official fisheries agencies would be used in combination with a LFC Micro-Credit Scheme to mobilise and disburse locally needed development and operational funds.

Loan assistance through the LFC Micro-Credit Scheme would in turn be used in combination with credit facilities to large commercial firms (linked with initiative 2 below), environmental education (linked with initiative 3 below), and new techniques of fresh fish collection and preservation to encourage:

(a) the replacement and retirement of destructive gear and fishing methods; the use of improved fish handling and processing methods to ensure better fish quality and hence the possibility of obtaining higher market values for fresh and cured products; and

(b) redeployment of some industrial fishing units to serve as collection vessels in order to reduce fishing pressure in southern waters.

2) Improved Infrastructure and Services

Activities foreseen under this initiative include:

(a) the technical support for improved local post-harvest practices;

(b) the provision of social facilities and services in pilot villages; and

(c) the installation of strategic marketing centre infrastructure and services.

Sun-drying on sand or gravel beds is the most common form of processing small pelagic clupeids ('sardines'), and this practice results in a product loaded with grit and other contaminants. The use of drying tables or concrete slabs, sometimes in conjunction with light brining and/or smoking, produces a cleaner, grit-free product known as '*dagaa safi*' (Kiswahili for 'clean sardines'). Brining and table drying of *L. stappersii* is also carried out in some localities.

Training and extension services to be provided under the programme are intended to increase the quantity of Tanganyika '*dagaa safi*' being sold at major retailing outlets both within and outside of the lake basin.

TREFIP grants-in-aid, subject to local co-payment or contribution in kind, will be provided for the construction or rehabilitation/upgrading of pilot village facilities e.g. schools, health centres/dispensaries, domestic water supplies and distribution networks, and latrines.

It is also proposed to improve links between important fisheries production centres and market places and to upgrade marketing facilities at key trading centres around the lake. Undertakings would include the construction or upgrading of selected roads, jetties, central markets, and electricity services. Provision is also made for a credit facility to fund establishment of fresh fish collection, handling, and marketing systems in certain areas. In Zambia, where marked overcapacity has developed in the industrial fishing sector, it is envisioned that this credit facility will be used to establish fresh fish collection operations using re-fitted purse seiners as transport vessels in combination with ice production units, chilled storage facilities, and new product evacuation channels. In the other countries, development of fresh fish collection and marketing systems would be encouraged through construction of small flake ice plants and chilled storage units.

3) Protection of stocks and biodiversity

Activities to strengthen protection of stocks and biodiversity include improvement of fisheries monitoring procedures (linked to initiative 5 below), establishment of lacustrine protected areas (PAs), and environmental education.

In the context of extensive consultation with stakeholder communities and agencies, potential sites for lacustrine PAs would be identified on the basis of a number of criteria. These include:

(a) existing status (proximity to national park/reserve boundaries);[39]
(b) importance as breeding and stock recruitment zones, and as aquatic habitat/bottom type sites supporting fish and invertebrate fauna with an exceptionally high degree of endemism; and
(c) accessibility, scenic setting, and the general likelihood of offering significant attraction as ecotourism destinations.

Six localities identified as *provisional* candidate PA sites include areas adjacent to national parks and major river deltas in Burundi, Tanzania, and Zambia.

Environmental Education (EE) is to be developed as an outreach programme through collaboration with LFCs, national fisheries researchers and managers, and NGO agencies involved with natural resource conservation and community welfare projects around the Lake Tanganyika littoral. Relying heavily on local language video presentations as well as posters, pamphlets, and community workshops involving extensive dialogue with fishers as 'indigenous environmental knowledge experts' (Makinson and Nøttestad, 1998), outreach activities would highlight the following topics:
(a) fishing and fish biology, lacustrine ecology and production system dynamics (e.g. importance of species diversity and protected areas, and impacts of different gear types and methods);
(b) post-harvest practices and the environment (impacts of fish smoking, reforestation needs);
(c) agricultural practices and water quality (soil erosion and sedimentation causes and effects, preventive measures);
(d) sanitation and health practices (causes and prevention of water-borne diseases and sexually transmitted diseases including HIV/AIDS; issues of family and reproductive health).

4) Improved fisheries legal regimes and MCS capabilities

TREFIP proposes to facilitate harmonisation of fisheries legislative frameworks and elaboration of specific regulatory measures for Lake Tanganyika, and the upgrading of MCS competencies along mutually agreed

---

[39] At Gombe Stream National Park in Tanzania, a buffer zone extending from the shoreline 100m into the lake already exists. Beach seining has been banned along the park shoreline since 1998, but three gill net licences are issued on an annual basis by the park authorities for fishing within the buffer zone area. At Mahale Mountains National Park in Tanzania and Nsumbu National Park in Zambia, the reserve boundaries extend 1.6 km into the lake. Fishing is banned within the reserve area of Mahale Park. Fishing is permitted from June to November at Chisansza Beach within Nsumbu Park, whose authorities issue licences and collect fees from annual users. It has been documented through LTBP studies that the restricted areas around the above three parks afford appreciable protection for fish species (West, 2001:45).

lines, through provision of technical assistance at both national and regional levels.

Particular attention would be given to securing provision for:

(a) the participation of stakeholders in management decision-making functions through CFMZs and LFCs (consultations to identify planning and development priorities and problem areas, elaborate regulations, etc.);
(b) new property rights regimes allocating control of access at the community level, in order to counter the 'race to fish' tendency that free access regimes entail; and
(c) the establishment and operation of enforcement and compliance assurance mechanisms under local responsibility.

5) More effective use of scientific advice for management

Since planning and management processes for Lake Tanganyika fisheries will be impossible to pursue in future unless a regular lake-wide monitoring programme is kept in place, TREFIP intends to provide technical assistance, training, and facility upgrades in order to build on the monitoring activities initiated under LTR and LTBP, further strengthen statistical capabilities within the respective national fisheries agencies responsible for Lake Tanganyika, and consolidate an institutionalised basis for co-operation among the respective agencies.

The existing LTFMP developed for national execution during the last stage of the LTR Project would provide the core elements of the revised monitoring system. However, provision would also be made to cover catch assessment and socio-economic parameters in data collection routines to be used within the newly established pilot CFMZs, in partnership with members of LFCs. In this connection full provision needs to be made for incorporating local people's 'expert knowledge' of their immediate terrestrial and aquatic environments (Makinson and Nøttestad, 1998). Further provision is also needed to cover biodiversity parameters according to techniques developed by the LTBP Biodiversity Special Study team for recording data on habitat characteristics as well as fish and mollusc communities at designated sites. A programme of in-service skill development for national staff and office and equipment upgrades would be initiated in order to ensure proper implementation of the expanded LTFMP.

6) Establishment of a regional fisheries management entity

In accordance with FFMP recommendations TREFIP would seek to establish a 'Lake Tanganyika Fisheries Centre' to:

(a) serve as a secretariat/executive arm for a Lake Tanganyika Regional Fisheries Council, to function as an umbrella organisation for LFCs and NFCs;

(b) facilitate technical investigations and discussions for all fisheries-related matters, including coastal zone management, and environment and water quality;
(c) promote exchange and dissemination of fisheries information, including operation of a Regional Fisheries Documentation Centre;
(d) develop, recommend, and facilitate implementation of conservation and management measures;
(e) in consultation with fisheries stakeholder groups, facilitate periodic review and revision as appropriate of the FFMP, taking into account the experiences and recommendations of TREFIP; and
(f) facilitate continued harmonisation of national policies and policy instruments pertaining to the sustainable utilisation of the living resources of the lake, in accordance with the CCRF principals embodied within the FFMP.

Exercising options for fishery futures

TREFIP has as its main objective the implementation of the FFMP. It has thus been framed in 'responsible fisheries' terms, and accordingly makes use of a holistic, or ecosystem-based perspective on management of fisheries systems and emphasises precautionary and participatory purposes.

The programme attempts to serve precautionary purposes in manifold ways. It does so directly by promoting the reduction of fishing effort level and the use of destructive fishing gear, and by promoting the establishment of Protected Areas, which would be constituted as 'no fishing' zones. It does so indirectly by promoting economic diversification, preservation of the basin environment through more sustainable exploitation of land and forest resources, and general appreciation of conservation values, including the need for balanced population growth, through environmental education undertakings.

Participatory purposes are to be served through the establishment of Community Fishing Zones and Local Fisheries Councils, National Fisheries Councils, and a Regional Fisheries Council. LFCs would be the primary tool through which small-scale fisherfolk will become involved in all aspects of management decision-making and implementation, including the allocation of resource access rights and responsibility for regulatory enforcement and compliance.

The initiative option: intended yields, inherent risks

Whilst TREFIP is intended to produce a wide array of positive impacts, it must be recognised that there are inherent risks of adverse impacts as well. Examples of possible benefits and risks associated directly or indirectly with programme activities are briefly outlined below.

1) Programme impacts on the lacustrine environment

The introduction of a micro-credit scheme is a key component of the TREFIP initiative to promote management partnerships around the lakeshore. For the new LFCs to become meaningful for local stakeholders, they will need to serve very practical functions and to become self-sufficient in their operations. Substantial benefits are likely to accrue as loan assistance is provided to help local fishing operators switch to more 'eco-friendly' gear, and to encourage local post-harvest operators in the use of handling and processing methods to obtain fresh and cured products of higher quality and market value. These measures, in combination with steps to redeploy a portion of the industrial purse-seining fleet into fresh fish collection activities, should contribute towards the reversal of trends towards over-exploitation.

There is some risk, however, that credit for input acquisitions will in localised instances draw new entrants, or former operators back into the fishery, thus countering efforts to reduce fishing pressure on over-exploited stocks. A related risk pertains to credits directed towards improvements in post-harvest handling and processing methods, and industrial fleet restructuring, and to efforts at infrastructure and service improvement. The intention is to foster more effective utilisation of harvests without concomitant increase in fishing effort. However, if significant rises in the price of fish and fish products should result, a nullifying trend of increasing fishing pressure and enticement of new entrants into the processing and trading business could ensue.

Steps to create lacustrine PAs under the 'protection of stocks and biodiversity' component of TREFIP involve complex and controversial issues. A large literature exists on protected or 'no take' areas and their role in serving conservation needs, particularly in regard to coral reefs (e.g. Allison *et al.*, 1998; Murray *et al.*, 1999). Observers have also commented on reserve areas with respect to the diverse zoo-benthic and fish communities (e.g. cichlids) in Lake Tanganyika (Lowe-McConnell, 1987; Cohen, 1991; Martens, *et al.* 1994).

The effectiveness of even the best-designed networks of PAs will depend on conservation and fishery management efforts undertaken outside reserve boundaries (Murray *et al.*, 1999). Individual reserves or reserve networks cannot alone produce desired fishery and conservation outcomes (Roberts, 1998). To become viable in practical management terms they require multi-level actions (Fogarty 1999; Murray *et al.*, 1999). In the context of the Lake Tanganyika region, this will include managerial actions taken simultaneously with respect to habitat protection, land-based operations, and monitoring and control of tourism and fishing activities.

Main production of the three commercially important pelagic species takes place in open waters. Of the clupeids, however, *L. miodon* spawns and hatches on sandy substrates in waters less than 130 m deep (Matthes, 1967), and produces large schools of juveniles (15-40 mm long) that spend their first two months inshore (Pearce, 1985c; Mannini 1998). The role of PAs in maintaining the commercial fish stocks whose life-history is only partly comprised of benthic- or littoral-dwelling stages will therefore be modest – particularly in the absence of measures to discourage the use of highly destructive beach seine gear.

Benefits will probably be more apparent in the form of sustained biodiversity of littoral dwelling cichlids. This is no small consideration, however, as the health of inshore and pelagic fisheries are closely interdependent. A collapse of the latter would precipitate significantly greater fishing pressure on the former. And marked depletion caused by over-fishing and habitat destruction within the littoral zone could easily lead to a transfer of fishing effort to the pelagic sector, already subject to over-exploitation risks of its own.

Also, to the extent that PAs will encourage the growth of ecotourism and thus economic diversification within the Tanganyika basin, they will benefit fisheries sustainability. The existence of employment alternatives to fishing would tend to retard the growth of exploitation pressure on commercial fish stocks.

A further point to be borne in mind is that PAs, besides directly benefiting exploited stocks, may provide an ecosystem-based management tool that focuses on processes and functioning, and extends fishery and conservation benefits beyond individual target populations (Roberts, 1998).

Adverse effects for the lacustrine environment attendant upon the establishment and operation of PAs can be expected to be negligible for the foreseeable future. The development of ecotourism, should it prove successful, is not anticipated to be of such a scale as to cause major negative impacts to aquatic habitats and fish. There is some risk that no benefits will be forthcoming at all from PAs, however, in the event that continued scarcity of alternative income-generating opportunities, and/or increases in the number of fishers, and/or lack of environmental awareness amongst local resource users would lead to violations of reserve boundaries and regulations.[40]

---

[40] Allison *et al.* (2001b) argue that there is little likelihood that profitable ecotourism will develop around Lake Tanganyika in the near future, and that the lake presents a case in which the benefits of integrated conservation and development programmes would probably accrue internationally (in the form of biodiversity conservation) whilst its costs would probably be borne locally (infrastructure development requirements, loss of opportunity for natural resource extraction, livelihood strategy options, etc.). There exists, as we note, a risk of 'no benefit' outcomes from ecotourism. On the other hand, should the very real challenges to its development be engaged, quite reasonable benefits could be realised. Moreover, as we have also noted, PAs would be likely to yield a very tangible gain in the form of better fishing – an outcome that would be most appreciated by local communities.

2) Programme impacts on the terrestrial environment

Management partnership arrangements are expected to serve a positive role vis-à-vis forestry and land-use practices around the lakeshore as they will provide community-based structures through which actions can be mounted to promote sustainable resource exploitation practices. Direct benefits would be forthcoming from improved fish processing operations encouraged by the micro-credit scheme and associated technical support. Reduced demand for fuel wood supplies would result from declining reliance on smoke curing of fish. Improved roads, jetty construction, provision of electricity supplies, and more efficient fresh fish collection, handling, and marketing networks also have a strong likelihood for favourable impacts in terms of reduced pressure on forest resources for the supply of wood to fuel fish smoking operations.

PAs ideally would encourage the growth of ecotourism and thus economic diversification within the Tanganyika basin, possibly leading to reduction in activities like woodcutting, charcoal burning, and brick production that exploit large quantities of forest and land resources. Greater awareness of environmental problems and ecosystem-based management approaches, nurtured through educational partnerships between local community residents and TREFIP government and NGO agency associates, would potentially have far-reaching positive effects in promoting sustainable forestry and land-use practices.

On the other hand, infrastructure and service improvements might promote increased commercial activity to such an extent that many more 'economic migrants' would be drawn to the Tanganyika littoral in search of employment and income opportunities. The opening of easier marketing routes for timber and other natural resources is also problematical. Such developments would be associated with further clearing of indigenous forest and bush areas for timber and fuel wood supplies, further agriculture encroachment, and further erosion and land degradation.[41]

3) Programme impacts on human welfare

Community survey work under LTR confirmed that, although views differ to some extent around the shoreline, local fisherfolk generally welcome the prospect of sharing management responsibilities. Participation in the pilot CFMZ/LFC programme will be voluntary, with residents of particular

---

[41] The intensification of smallholder cultivation and other land use impacts within the Lake Tanganyika catchment in recent decades was the subject of detailed investigation under the LTBP, through its study on sediment discharge and its consequences. This study concluded that a dramatic five- to tenfold increase in the input of suspended sediments has occurred from the pre-1960s period. Littoral sites within 10 km of the point of discharge of a medium-sized (40 – 50,000 km^2) catchment area appeared to be most threatened by erosion within that area (Patterson, 2000).

communities deciding amongst themselves whether they want to join in. It is anticipated that the programme will foster a strong sense of solidarity and civic purpose amongst those who elect to participate. Such qualities would be conducive to the development of community self-help initiatives in areas beyond the affairs of fishing.

Possible adverse effects include disproportionate empowerment of various segments of pilot village community populations. Women, though very active in fisheries post-harvest activities, tend not to share the advantages of educational attainment and income earnings realised by their male fisher or processor/trader counterparts. Women's comparatively low profile in civic affairs is also disproportionate to the socio-economic contribution they make as fish workers, businesspeople, farmers, and household providers.

Others involved in fisheries-related work, especially as crew and helpers for more affluent boat and gear owners, similarly are often subject to the disadvantages of subordinate social status and poor income levels. If women and low-income members of pilot communities are excluded from meaningful representation in LFC structures, such marginalisation would simply be reinforced and the implementation of the co-management mechanisms proposed under TREFIP would thus tend to run counter to responsible fisheries goals of socio-economic equity.

Substantial community benefits can be foreseen as a result of component activities related to micro-credit, introduction of appropriate gear, and fleet restructuring. Local fisherfolk face great obstacles in seeking loans from established commercial banks. Operation of the LFC Micro-credit Scheme will enable them to upgrade or modify their productive equipment and techniques in ways that not only will foster fisheries sustainability; it may lead to higher market values for fresh and processed fish as well, and thus to enhanced earning levels and living standards.

Risks include those already noted above – namely, the chance that credit for input acquisitions and fleet restructuring for more efficient fish collection might: a) attract new entrants into, or former operators back into, the fishery; or b) promote rises in the price of fish and fish products that would in turn increase demand for raw product. Such developments would run counter to aims of fishing pressure reduction on over-exploited stocks.

A further possible adverse effect would arise in the event that women and low-income fish workers and processors/traders were to be excluded from, or not adequately represented, in the company of those receiving micro-credit assistance.

Very considerable positive effects can be anticipated from improved post-harvest practices. Local processors would benefit from reduced drying times in the treatment of small pelagics, and both processors and traders would benefit from extended product shelf life, and, probably, better product prices. Consumers, whether local or remote, would benefit from a cleaner and healthier product. Finally, local communities and fish consumers in urban centres and areas distant from Lake Tanganyika stand to benefit from a greater supply of fish because of reduced wastage from breakage, fragmentation, and spoiling of product as it passes through distribution and marketing channels.

Possible risks lie, once again, with the indirect effects that may arise from improved terms of trade, such as attracting new entrants into the fishery. Similar mixes of quality of life gains and adverse impact, particularly in terms of 'economic migrant' attraction, are associated with TREFIP community partnerships for improvement of physical infrastructure and social services.

Establishment of lacustrine protected areas will yield advantages to adjacent communities in the form of minor short-term employment opportunities during facility installation phases. Once operational, PAs will offer long-term employment opportunities in proportion to the amount of ecotourism traffic that is attracted. Economic benefit to local communities will also depend on managerial organisation. As learned from experiences with national parks and game reserves in other parts of Sub-Saharan Africa, community-based management and local ownership of park areas and tourist facilities fosters strong employee commitment. Local worker-owners are eager to embrace proper management techniques and the better economic returns they yield, as opposed to employees of organisations controlled by remote parastatal administrations (cf. Balakrishnan and Ndhlovu, 1992; Murindagamo, 1992; Siachoono, 1995). The new LFCs could play an important role in developing management and services within PAs.

Improved legal regimes and MCS competencies would, amongst other things, empower local stakeholder groups, through CFMZs and LFCs, as direct participants in fisheries management decision-making and as principal responsible parties in enforcement and compliance assurance activities.

Once LFCs are established as formal entities, they would serve a positive role as focal groups for the deliberation and amelioration of environmental problems with which Tanganyika fishing communities are often confronted. These problems, complex and interrelated consequences of expanding populations, unsustainable fishing and agricultural practices, and deforestation and land degradation, are ultimately best dealt with through concerted actions by members of the affected communities themselves.

There is a risk, however, that termination of open access conditions for the Tanganyika fisheries and allocation of access restriction responsibility to local stakeholder groups will prove problematical in the absence of viable local or regional employment alternatives for those who will no longer be able to recruit themselves to the fisheries.

An expanded LTFMP would provide information on socio-economic parameters of the Tanganyika fishery, including changes in settlement population size and composition, infrastructure, and amenity features of lakeshore communities. Monitoring would thus serve as a tool for gauging the capacity of the natural environment and social infrastructure to cater for human welfare needs.

The Tanganyika Regional Fisheries Council and its secretariat, the Tanganyika Fisheries Centre, will provide mechanisms for the compilation, analysis, and dissemination of fisheries information on a lake-wide scale, and for the co-ordination of technical measures needed to ensure the sustainable use of resources.

Such measures would be informed of the actions and recommendations of LFCs, and of the routine socio-economic monitoring work conducted as part of the LTFMP. Community interests would therefore be strongly represented in efforts to further responsible fisheries management processes on a regional basis.

The default option: doing nothing

Exercising the 'initiative option' will obviously require mobilisation of considerable resources in terms of finance and political will at local, national, regional, and international levels. It will also require far closer collaboration between major implementing and executing agencies involved with Lake Tanganyika fisheries projects than has been achieved heretofore.[42] Above all, it will require a favourable 'enabling environment' in terms of regional security – a condition that seems very remote of attainment insofar as some areas are concerned. The East-Central zone of Africa remains an arena of vicious conflict, pillage, and population displacement. The civil war in Burundi continues along its tragic course, and circumstances in war-torn DR Congo remain in a state of great uncertainty.[43]

---

[42] This especially includes the various UN agencies that have been associated with recent projects on the lake. Although some collaborative work has been carried out between them, LTR (FAO) and LTBP (UNDP/GEF) have often operated as if their purposes were competitive rather than complementary. The LTBP SAP was prepared not only in a largely 'top-down' fashion by planners and decision-makers with little contact with the day-to-day realities of the lake. It was also prepared with scant reference to the substantial body of LTR research reports on various hydrobiological, socio-economic, and legal/institutional aspects of the commercial fisheries, and with little effort to include LTR team representatives, including counterparts from national fisheries agencies, directly in drafting deliberations.

[43] In mid-2001 hopes arose that a ceasefire between the various factions, the partial withdrawal of some foreign occupation forces and the inception of a UN peace-keeping mission could represent the initial steps down what is bound to be a very long path to the recovery of stability and civil society in the DRC.

In the face of such discouraging prospects, it may be tempting to fall back on the 'default' option for Lake Tanganyika fisheries initiatives – i.e. to put them on indefinite hold. Yet such a 'nothing can be done for now' response entails an even more discouraging prospect for fisheries futures on Lake Tanganyika – one that is truly bleak.

Without the kinds of initiatives proposed under TREFIP, management and conservation of the natural resources and fish stocks of Lake Tanganyika would rely on the continuation of the present medium of co-operation between the four lacustrine countries, and the FAO-facilitated Lake Tanganyika CIFA-Sub-Committee. The operations of this body are intermittent (meetings once every two years, at most) and certainly not geared to engage the comprehensive and immediate management challenges presented by the lake and its fisheries. Moreover, the Sub-Committee is heavily dependent on external (FAO) financing, which is not likely to continue indefinitely. Recognising these weaknesses, the Sub-Committee at its 8th Meeting (Lusaka, May 1999) deliberated and adopted new Terms of Reference. These TORs stipulate that the Sub-Committee should, amongst other things, facilitate the creation of a permanent regional fisheries management body for the lake.

On a country level, the 'do nothing' option would imply continuation of present arrangements under which the respective national fisheries administrative and research agencies and their field personnel are responsible for management decision-making, including technical review and advice, MCS, and enforcement. Based on experiences thus far, budgetary, technical and organisational deficiencies impose severe constraints on the ability of these agencies to fulfil their responsibilities either generally, i.e. on a country-wide basis, or specifically in connection with the remote (except for Burundi) fisheries of Lake Tanganyika. In the absence of the kinds of initiatives proposed under TREFIP, there is little likelihood that various adverse trends affecting the lake's fisheries can be arrested, let alone reversed. Such trends include the following:

(a) growing exploitation rate of both demersal and pelagic fish stocks due to increased demand for fish (burgeoning human populations) and limited control measures;
(b) increased risk of destroying sparse populations of *L. mariae*, *L. microlepis*, and *L. angustifrons*;
(c) growth and recruitment over-fishing of *L. stappersii* and *S. tanganyicae* due to industrial and advanced artisanal operations;
(d) heavy damage inflicted on littoral-borne life stages of pelagic species due to uncontrolled beach seining and use of other harmful gear;
(e) reduced unit catch and economic return;

(f) loss of fish quality and quantity through the post-harvest handling stages (processing, transport, and marketing), including increased risk of disease due to poor hygiene and insufficient facilities;
(g) weakened nutritional and health status amongst children and urban fish consumers; and
(h) continued degradation of the lacustrine and terrestrial environment and loss of productive capacity of the basin ecosystem.

The utilisation level of fishery resources is presently very high, and any uncontrolled increase of exploitation will render the ecosystem and fish production more susceptible to rapid environmental changes. This evolution would lead into increased ecological uncertainty, which, in turn, would add to economic uncertainty for fishers.

A future fraught

The 'do nothing' option thus by no means implies a 'no impact' outcome. What it amounts to instead is a recipe for further and probably accelerated environmental and socio-economic deterioration. It is, in other words, no option at all if responsibility is to figure in fishery futures for Lake Tanganyika. And these futures are upon us. Even if sections of the western lakeshore remain out of bounds until warring factions lay down their arms and a measure of security and civil society can be established, there is no reason to delay undertaking TREFIP initiatives elsewhere.

These initiatives should furthermore not be held in abeyance whilst follow-up work on the GEF Strategic Action Programme and promotion of a Lake Tanganyika Convention and a Lake Tanganyika Authority proceeds on its protracted and somewhat indeterminate course.[44] Whilst the SAP and its institutional mechanisms may all make ultimate sense as the best way to achieve regional multi-sectoral management of resources within the Tanganyika basin, putting them into meaningful effect is bound to entail many years of effort. The vision may be an excellent one, though it will need a genuine addition of local stakeholder and fisheries perspectives as well as political accommodation at the highest levels to help bring it into effective focus. Regarded realistically, this focusing process will be at least decadal in scope.

The fisheries of Lake Tanganyika are in too much trouble and are too important as a basic engine of regional livelihoods for their management to be kept pending, held as a kind of hostage in some planning limbo (cf. Coulter,

---

[44] As previously noted, UNOPS and UNDP/GEF are presently engaged in a one-year interim planning exercise to formulate projects that would support implementation of the Strategic Action Programme. However, collaboration between FAO, UNOPS, UNDP/GEF and the AfDB remains largely of a nominal order.

1999), as we wait for this vision to resolve itself. Immediate mobilisation for responsible fisheries on the lake, including the leveraging of all necessary donor support, would provide a solid basis for eventual elaboration of a more comprehensive, basin-wide, multi-sector resource conservation and management programme and institutional structure. Waiting for the basin programme and institutional structure to be elaborated first, and for the enabling Convention to come into force, seems a little like falling back on the default option for the lake and its fisheries – saying, in effect, that 'nothing can be done for now.'

Although we conclude this contribution on what is perhaps a less than upbeat note, we do not want to be unduly pessimistic. The FFMP and its implementation programme, TREFIP, obviously do not propose easy and totally guaranteed 'fixes' for the multiple fisheries challenges confronting Lake Tanganyika. It would be gratifying to offer such a package - a simple, straightforward, and effective management partnership scheme immediately acceptable to all local resource users as well as the various national and international parties – fisheries administrators, scientists, conservationists, planners, and policy makers - who share concerns for the well-being of the lake. The FFMP and TREFIP proposals, some might complain, are nowhere close to this mark. We can only say in response that we believe it naïve to expect that such a ravel of management challenges as Lake Tanganyika presents can be dealt with in a way that assures neat, speedy, and unconditional success. We maintain that these challenges can only be addressed on a processual basis, in conformance with responsible fisheries principles that call for careful deliberation with stakeholders, prudent assessment of risk, and a willingness to engage in adaptive, trial-and-error adjustments to management circumstances. Opting out of these challenges altogether may seem to offer the ultimate in ease and simplicity. What it really involves, we reiterate, is the ultimate in risk. The more responsible way forward will require proactive yet precautionary engagement of the complex of potentialities, constraints, frustrations, uncertainties, and doubts that Tanganyika poses for its users and keepers. We know the way forward will not be easy. It will be a future much fraught.

Acknowledgements

The authors would first of all like to pay tribute to the fisherfolk of Tanganyika, who have consistently provided us with warm hospitality and patiently shared with us their insights on the lake and its transitions over nearly a decade of LTR project activities. They well know, above all of those who have experienced the realities of the lake and are concerned with its future, that the gap between what can be wished for or prescribed and what can actually be effected is often difficult to bridge. We also gratefully acknowledge the contributions of the many colleagues with whom we have shared both hardship and fulfilment during the course of LTR, including most especially the scientists and officers affiliated with the project as staff of the national lakeside fisheries stations in Bujumbura, Uvira, Kigoma, and Mpulungu. Our thanks are also due to the international scientists and advisors who have participated in LTR from the FAO, and from the Universities of Kuopio, Turku, Jyväskylä, Joensuu and Helsinki, the Game and Fisheries Research Institute, and the Environmental Agency in Finland. George Hanek, FAO resident LTR Coordinator from 1992 through 1999, deserves a special vote of thanks for his energetic efforts to ensure successful project operations in what were frequently trying and frustrating circumstances. The practical assistance and suggestions often provided by our colleagues in the LTBP is also gratefully acknowledged. We furthermore express our appreciation to the editors of this volume and to the anonymous reviewers who read an earlier draft of this article and offered useful comments, which we were able to take into consideration in preparing the final manuscript. It goes without saying that the views and opinions expressed in the article are those of the authors alone, and do not necessarily reflect the views of FAO, the University of Kuopio, or the Government of Finland.

8
On Pitfalls and Building Blocks: Towards the Management of Lake Victoria's Fisheries

Kim Geheb, Kevin Crean, Modesta Medard, Mercy Kyangwa, Carolyne Lwenya, and Paul Onyango

This paper sets out to explore the management of Lake Victoria's fisheries. It argues that the main problem in the lake's management stems from three inter-related 'meta-problems':

(a) In Kenya, the failure to consult with, or include, fishing communities at any point of the management structure has severely undermined the regulation of the fishery, particularly when coupled with poor regulatory implementation and enforcement by the Fisheries Department.

(b) In Tanzania, new regulatory initiatives that seek to use communities to implement the centralised management system are crippled by the often repeated concern that where fishing communities are not involved in the design of the regulations they are supposed to adhere to, it is unlikely that regulation will occur (Johannes, 1981; Ostrom, 1996). Instead, this new management structure has been absorbed by fishing communities and 'socialised' in such a way that it can be accommodated in wider attempts by communities to meet livelihood objectives (Medard and Geheb, 2000). This accommodation might not yield a managed fishery.

(c) In Uganda, government efforts to transfer decision-making and implementation powers to local communities are laudable. Fishing communities have not, however, seized this initiative. In addition, there are gaps in the legislation, for example, the actual roles of stakeholders within the new administration are not identified.

These problems manifest themselves in different ways and give rise to a series of dilemmas. These are explored in the first part of this paper. In the second part, we suggest possible solutions to these problems, and propose a partial managerial model based on these solutions. The model is dynamic and multi-layered, and presents opportunities for both the decentralisation and devolution of power to lake-side communities. It proposes redefinitions of the roles of various actors, the legal endorsement of the structure, the 'nesting' of the various levels within a broader regulatory framework, and an integrated system of regulatory sanctions and justice. At the local level, prescriptions are minimised. As the organisation of the plan becomes more formalised and includes higher degrees of government and private sector involvement, so too prescriptions increase to meet the demands of high levels of organisation. We do not pretend that the plan proposed is a complete management plan, but hope that the suggestions made in this paper represent the way forward for the development of a unified, lake-wide management plan that offers greater potential for success than present managerial strategies. The structures presented, we argue, will yield the best possible results given the situation that the fishery presently finds itself in. These may not be as good as demanded by management, but represent a negotiated compromise between the livelihood claims of fishing communities and formal management objectives.

The data and experiences on which this chapter draws, unless otherwise referenced, are all derived from research conducted under the Lake Victoria Fisheries Research Project (Geheb, 2000b; Geheb and Crean, 2000; Geheb *et al.*, 2000a, 2000b, 2000c; SEDAWOG, 1999a, 1999b, 2000)

Lake Victoria

Lake Victoria is a massive inland water covering 68,000 km^2 and lying across the borders of three countries: Kenya, Uganda and Tanzania (see Figure 18).

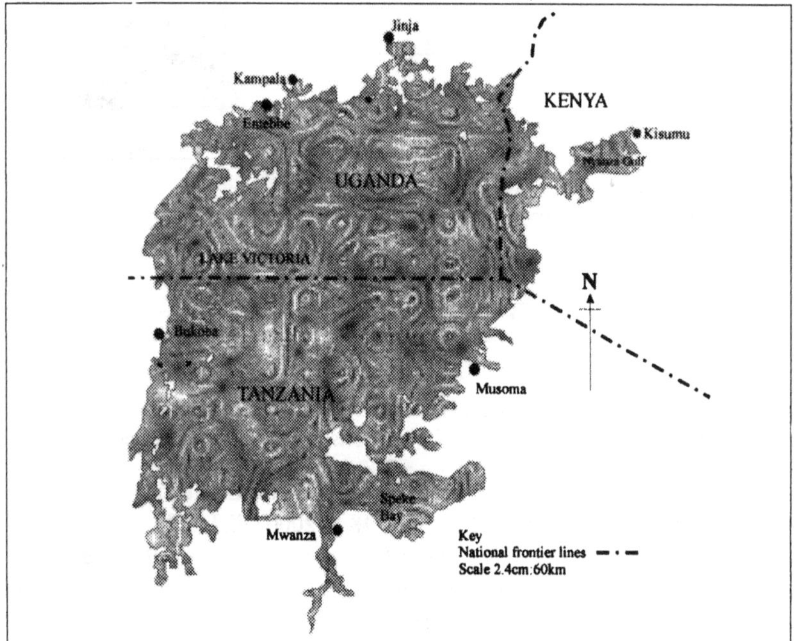

Figure 18: Lake Victoria

The principal ethnic groups inhabiting the lake margins are, in Uganda, the Baganda, the Basoga and the Samia; in Kenya, the Luhya, Luo and AbaSuba; and in Tanzania, the Sukuma, Haya, Ukerewe and the Mjita (Geheb *et al.*, 2000a). In the past, the lake teemed with several hundred species of fish, most of which belonged to an extended species flock, the haplochromines. In the early 1980s, following the population explosion of the introduced predator, the Nile perch (*Lates niloticus*), haplochromine populations crashed, resulting in what may be the biggest mass extinction event in modern history (Goldschmidt *et al.* 1993; Witte *et al.* 1992). The fishery is presently dominated by three commercial species: the introduced Nile perch and Nile tilapia (*Oreochromis niloticus*), and the endemic '*dagaa*', a small pelagic cyprinid (*Rastrineobola argentea*). A total of 117,757 t of fish were landed from the lake in 1968. Following the Nile perch population explosion in the 1980s, landings peaked in 1990 at 787,899 t. Since then, catches have declined by 48% and in 1995 stood at 406,799 t (see Figure 19).

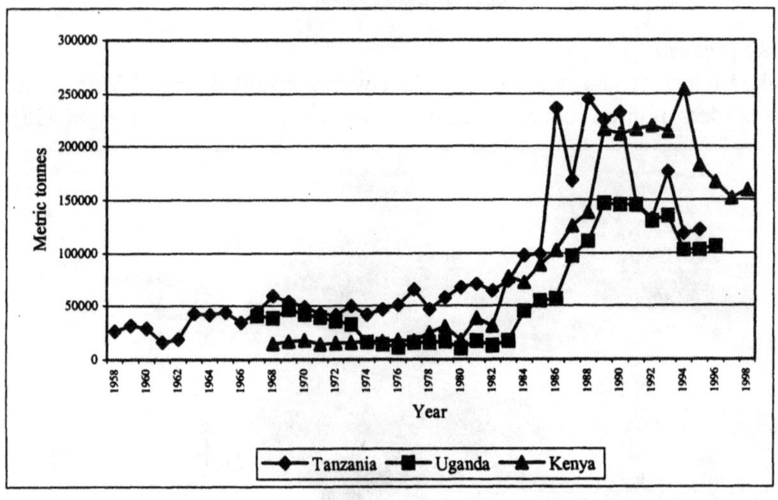

Figure 19: Total catches from Lake Victoria, 1958 - 1998. **Sources:** Greboval and Fryd, 1993; Reynolds and Greboval 1988; unpublished fisheries department statistics.

Levels of fishing effort have grown exponentially, from an estimated 12,041 boats in 1983, to 22,700 in 1990 and 42,548 in 2000 (FSTC, 2001; Greboval and Fryd, 1993; Hoekstra *et al.*, 1991; Mkumbo and Cowx, 1999). Besides the income that accrues to the region's fishermen, the fishery is also an important source of foreign exchange. In 1996, Kenya exported 16,472 t of

Nile perch fillets abroad, while Tanzania exported 15,000 t and Uganda 16,396 t (Lake Victoria Fisheries Organization, unpublished data). In 1996, the value of Uganda's export of fish products was US$ 46,251,000, second only to coffee (RoU, 1999).

Background to the regulation of Lake Victoria's fisheries

Following the establishment of the Lake Victoria Fisheries Service (LVFS) in 1947, the regulation of the lake's fisheries has been dominated by 'command-and-control' styles of fisheries management, under which fishing communities are supposed to adhere to various fisheries regulations on pain of punishment if disobeyed. Declining catches, increasing levels of effort and other evidence (cf. Geheb, 1997; Reidmiller, 1994) suggests that this management style has not worked on Lake Victoria. Discussion within the region has, therefore, turned to considering possible managerial alternatives, chief amongst which is the concept of 'co-management'.

The term 'co-management' is ambiguous because it can promise a great deal without really specifying anything in particular. In much the same way that 'sustainability' before it became a new jargon term in the annals of development (Adams, 1990), so too co-management has become a necessary catch phrase within the realm of resource management. Broadly speaking, writers on the subject agree that co-management represents some sort of collaborative arrangement between the state and a community of resource users aiming to conserve a resource base (cf. Noble, 2000; Sen and Nielsen; 1996). In an era where flexible management strategies are called for to regulate dynamic resource use (cf. Ludwig *et al.* 1993; Walters and Holling, 1990), the ambiguity of 'co-management' allows managers to apply the necessary flexibility required for satisfactory managerial outcomes to occur. At the same time, the ambiguity of co-management can assist fisheries managers to design location-specific management programmes. A lack of dynamism and assumption of ubiquity, indeed, have been amongst some of the main criticisms of contemporary styles of fisheries management (cf. Crean and Geheb, 2001; Dustin Becker and Ostrom, 1995; Hilborn *et al.* 1995; Keohane and Ostrom, 1995). It is such a style of regulation that persists in Kenya. In Tanzania and Uganda, however, efforts have been made to alter management styles, although whether or not the changes that have occurred represent 'co-management' is debateable.

In Uganda, following the introduction of its new Constitution in 1995, the state guaranteed the decentralisation of many of its powers to district and lower levels of formal administration. The lowest of these levels is the Local Council 1 (LC1), which operates at the ward level, and draws its representation from its inhabitants (RoU, 1995). In 1995, the Ugandan Government launched the Local Governments Act (*sic.*) (RoU, 1997) which further articulated concepts promulgated in the Constitution, and defined the procedures by which LCs could establish locationally-specific by-laws for all manner of issues and areas of local importance, and which can include the governance of resources.

The formal administration of Uganda's fisheries falls under the 1964 Fish and Crocodiles Act, subsequently re-named the 1964 Fish Act (Uganda Government, 1964). The basic precept governing the design of this Act was that it should remain fairly broad, so that fisheries administrators could issue government 'legal notices' as and when the need arose – not dissimilar, in fact, to an adaptive management strategy (Walters and Hilborn, 1978). Hence, conventional fisheries regulation facets, such as minimum mesh sizes, or the prohibition of certain fishing gears, are not actually stipulated under the Act. Because the 1964 Fish Act is so ambiguous, room exists for fishing communities to formulate more or less any fisheries regulations they desire, provided these do not violate the laws of Uganda (Geheb, 2000a). By contrast, in Tanzania's Lake Victoria fishery, a more prescriptive approach has been adopted. Here Beach Management Units (BMUs) have been established, drawing their membership from fishing community residents (Hoza and Mahatane, 1998). BMUs are charged with enforcing and implementing the state's fisheries regulations.

The pitfalls

Regulatory ambiguity

Dustin Becker and Ostrom (1995) and Keohane and Ostrom (1995) are concerned about the ubiquitous application of management in many contemporary styles of natural resource management. Their main concerns relate to the fact that these management styles assume that fishing communities are homogenous, and such management systems cannot therefore cope with the many cultures, claims, contentions and access differences, within and between, communities of resource users. These problems are exacerbated when a resource – like Lake Victoria's fisheries – is a multi-species one. This is, of course, a design flaw, and would appear to be true no matter where in the world these systems are applied. In Africa, however, fisheries regulations, if at all enforced, are not usually ubiquitously applied. In many African fisheries, wealthy fishermen can pay to be overlooked by fisheries departments, while those who cannot afford such graft are punished (cf. Aarninck, 1999). This is one way in which regulations become blurred in Africa's fisheries. There are other ways in which this might be engineered, as we shall see below.

In Kenya, the structure of fisheries administration has not substantially changed since the establishment of the Fisheries Department in the early 1960s. Fishing communities have no room for participation, and low-level Fisheries 'scouts' will use the fishing communities' lack of knowledge about fishing regulations as a means to extort graft (Geheb, 1997). In the maintenance of this ambiguity, knowledge of the true regulations may even disappear. In one survey (Geheb *et al.*, 2000b), out of five Kenyan fisheries officers interviewed, three did not know what the minimum mesh size for gill nets was. The same survey also revealed that 52% of fishermen questioned did not know what the minimum mesh size for gill nets was (Geheb *et al.*, 2000a). The gaps between the fisheries regulations, what fishermen know about them and what fisheries department officials claim they are, present management with many problems. At one level, this ambiguity ensures that managerial outcomes do not even approximate to those anticipated. At another, the ambiguity can be used for reasons entirely different from what the regulations were originally designed for.

In Uganda, the Department for Fisheries Resources (DFR) has been restructured. Under the new scheme, DFR headquarters in Entebbe has responsibility for law enforcement, and relies on a Fisheries Regulations and Control Unit (FRCU). In 1998, the FRCU had a staff of 35 officers charged with maintaining fisheries regulations on all of Uganda's water bodies (of which there are 43,941 km^2) (Kiiza, 1998). At the same time, the formal hierarchy established under the 1997 Local Governments Act has created fisheries regulation systems at district and ward levels. At the most local level, fish 'guards' (equivalent to Kenya's Fisheries 'scouts') patrol between two or three beaches, and are answerable to fisheries officers at district headquarters. The main activities of the fish guards seem to revolve around income collection for local authorities, primarily through fish licence and boat registration fees. What their law enforcement role is, however, remains ambiguous, and for practical purposes, so long as there are fisheries personnel with ambiguously defined roles visiting landing sites, the impetus for fishing communities to fulfil regulatory roles must be seen as limited. In the absence of clearly stated responsibilities for each of these two groups of stakeholders, there is a very real danger that the opportunities created by the 1997 Local Governments Act for community-created regulatory by-laws will not be seized, because communities are unaware that they can seize them and do not know what responsibilities they may assume under the Act.

The Tanzanian approach to 'co-management' through the establishment of BMUs does not, it has been argued, represent the creation of a co-managerial relationship, but, rather, the subjugation of Tanzanian fishing communities to uphold the state's laws (cf. Geheb, 2000a; Medard and Geheb, 2000). Fishing communities see themselves as enforcing state regulations on behalf of the state, and hence, as doing the state's work. As a result, they demand remuneration. In the same way that other Tanzanian government structures have been imposed on communities, BMUs are likely to become 'socialised' into community structures, and will render regulations ambiguous. While this may not be a problem in some ways, in other ways it is, insofar as the ambiguity ensures that managerial objectives are in no way met.

The dichotomy between the levels of the state and community

Co-management implies power sharing, and a shift in power away from the state to the community. The nature of this shift is both in terms of power magnitude and quality, which directly impinges on the discretionary powers retained by the state. Commenting on the judiciary's reaction to the establishment of local-level *sungusungu* vigilante groups aimed at improving law enforcement in the Sukuma and Nyamwezi areas of north-western Tanzania, Bukarara argues that:

> The police and the judiciary...are unhappy with and opposed to Sungusungu taking the law in their own hands and providing an additional and/or alternative means of social control. Officials of these institutions argue that Sungusungu members are attempting to turn the clock back to primitive punitive measures...*The competition between the two suggests that Tanzanian state institutions are more concerned with the protection of the legality of monopoly of their powers than with the actual problem of crime* (Bukarara, 1996: 264. Emphasis added).

This problem of trying to retain monopoly over certain government powers prompts the question as to what, in fact, the management requirement actually is. Is it the retention of administrative bureaucracies for the sake of their members, or their retention for the sake of, in this case, fisheries management? In the absence of any transfer of powers to fishing communities, community-level powers do not exist to counter-balance excesses in centralised powers. Where regulations are ambiguous, then room exists for the powers associated with these regulations to be abused, and utilised for ends for which they were not designed (Chapman, 1989). It goes without saying, however, that a shift of administrative powers away from the centre to the periphery implies explicit faith in communities to carry out fisheries (and other resource) management roles. This difficulty is particularly acute in Kenya, where no attempt has as yet been made to involve communities in the development of fisheries regulations, let alone granting them the power to design these. In the absence of this power, the kinds of ambiguity described above become accentuated, and management objectives are undermined.

The devolution of administrative powers to Tanzania's fishing communities does not indicate faith in the abilities of fishing communities to design and implement their own fisheries regulations. The fact that communities have to implement government regulations, does, however, imply a belief that communities can be trusted to carry out these roles. This faith may, however, be misplaced given that government has been loath to invest resources to bring communities on board in terms of consultation, briefings, education etc.

Endorsement by the state of local-level institutions

The dilemma of the state-local level dichotomy also manifests itself in the problem of endorsing local institutions. Lake Victoria, like many African waters, is held in trust by the state, and fisheries departments are agents of government. As discussed earlier, fisheries departments may, on the one hand, fail to deliver regulatory outputs, while, on the other hand, deliberately work against fishing communities' attempts to take regulatory action because of the threat that this implies to their discretionary powers.

Semi-formal institutions exist and carry out relevant management activities at most of Lake Victoria's landing sites. In Kenya, beach committees collect landing fees from migrant fishers, which may be utilised for the management of the beach as a whole. In Uganda, following the widespread use of poisons to kill fish, the government closed the fishery, and created a series of 'task forces' to oversee the prohibition of fishing with poisons at the local level. Now that poisons are rarely used in the fishery, these task forces have evolved into Landing Management Committees (LMCs: Kyangwa and Geheb, 2000), which have various administrative tasks at the local level. In Tanzania, the Beach Management Units operate at the landing sites, along with other community-level administrative organisations, such as the so-called '*sungusungu*' vigilante groups, which are a part of Village Committees. All of these organisations have tremendous managerial potential within the fishery. It is in their limited powers that the dilemma exists. The nature of this deficiency, however, varies from country to country.

In Kenya, the extent to which local level institutions are permitted to exert control over fisheries activities is very limited. Geheb (1997) describes a case from Kaloka beach, where leaders complained to the Fisheries Department about lack of controls over fishing gear. The Fisheries Department responded that they should create their own methods of controlling the problem, and fishing community leaders decided amongst themselves that the local sub-clan would ban illegal fishing gear. They gained support from sub-clan elders and the area Assistant Chief. The Division Chief, however, intervened, saying that the local sub-clan could not take the law into their own hands, and that their plan would have to go through the 'proper channels'.

At Uhanya beach in 1993, fishermen disillusioned with the Yimbo Fishermen's Co-operative Society (YFCS) attempted to form a parallel organisation, the Uhanya Fishermen Organisation. Fishermen started handing their commissions over to the newly formed group rather than the YFCS. The YFCS then complained to the District Commissioner, who ordered the group to desist and that its members hand their commissions over to the YFCS (Mitullah, 1998: 4). Mitullah (1998) argues that such political pressure against the formation of fishermen's association – with both economic and administrative powers – ensures that such groups can only really perform 'welfare' functions.

In Tanzania, however, Medard's (2000) discussion of a small women's group based in Muleba District suggests that the creation of such groups at the welfare level, and their subsequent strengthening through their own internal activities, may create the necessary foundations for the attainment of economic powers. By initially establishing a group with fairly modest aspirations, the Tweyambe Fishing Enterprise (TFE) ensured that it in no way offended external political sensibilities. Its gradual assent to considerable economic importance within the community also occurred in such a way that the dominant political economy within the Kagera Region's formal administration was not upset. What is noticeable about the TFE is its marked departure from contemporary Tanzanian village structures imposed from beyond the community. This, indeed, may be part of the reason why the TFE has been so successful. What appears to characterise the success of the group is the convergence between its individual members' livelihood aspirations and the group's overall objectives. In other words, the TFE manages to ensure that there is a convergence between individual self-interest and community interests (Oye and Maxwell, 1995), without unduly upsetting the *status quo* of the dominant political economy within their area.

The latter cannot be said for externally imposed community structures elsewhere along Tanzania's lakeshores. The BMUs have been described above, and the difficulties that these face lie in the absence of community-level contributions to the regulations that the BMUs are charged with enforcing[45]. While it is clear that they do ensure that there is community participation within the management of Tanzania's Lake Victoria fishery, communities do not 'own' the managerial process (Onyango, 2000). As Wilson *et al.* (1996: 9) comment, 'The experts tell the fishers what they should know and what they will do, and the fishers participate [in management] by complying'.

---

[45] It should be noted that the Tanzanian Fisheries Department does consider the lack of legal backing of the BMUs to be a problem, in particular, the BMUs' lack of legal power to prosecute offenders under the 1970 Fish Act (Mahatane, pers. comm.).

In the process of 'socialising' these regulations, the objectives of the BMUs become ones of balancing their administration between the demands of the Fisheries Department and their own, localised, livelihood claims, without necessarily administrating the fishery *per se*.

The lessons of Kenya's and Tanzania's local level fisheries administration are that if such institutions are to evolve under present political structures, then they must do so without offending them. The very threat that they will cause such offence ensures that the state is reluctant to see the devolution of powers to these communities[46]. At the same time, by persistently laying claim to the responsibility for regulating the fishery, and excluding local communities from this process, the state ensures that communities never see responsibility for the resource base as their own. In fact, this problem contributes to the managerial impasse on Lake Victoria, for by retaining managerial responsibility for the lake and doing little to fulfil this role, and by denying (or preventing) communities from contributing to management, the government is likely to cause the creation of a vacuum in regulatory responsibility.

In Uganda, almost the opposite is the case. Here, the 1997 Local Governments Act demonstrates the state's willingness to see a significant devolution of powers away from the centre to the periphery. The difficulty lies in persuading fishing communities to seize the opportunities available to them. It is possible, however, that some communities do sense what these comprise. Following the abatement of poisoning in Uganda's fishery in 1999, the Government announced that the fishery was once again open. Several districts, however, refused to open their fisheries, claiming that fish poisoning was not yet sufficiently controlled. Faced with the potential that they might once again be hurt financially, many Ugandan fishermen felt that the fishery should only be reopened once poisoning had been completely eradicated. Provided, then, that the stakes are in fishermen's favour, then concerted action may occur – hence, with their income severely curtailed by the ban on fishing, fishermen responded to the ban on poisoning rapidly and harshly[47] and appear to have been determined to stamp it out. This action did not, however, translate into the widespread implementation of fisheries-related by-laws, and a management plan must not only deliver such rights to fishing communities, but also have provision to encourage fishing communities to exploit these rights.

---

[46] Noble (2000) makes the important distinction between the 'devolution' of power (which implies a shift of qualitative and discretionary power away from the centre to the periphery), and the 'devolvement' of power (which implies a shift of administrative responsibility away from the centre to the periphery). The distinction is of particular use in Tanzania's case.

[47] Ugandan fishermen were reported to have bludgeoned to death captured poisoners.

The issue raised by Hara *et al.* (Chapter 3) that co-management in many respects is an off-shoot of (often) donor-driven demands for democratic structures to be implemented in many of Africa's countries, is pertinent here. While the Ugandan and Tanzanian states may well be willing to devolve power away from the centre to the community level, fishing communities may not be ready – or even willing – to assume their rights under these kinds of systems, given the relatively authoritarian structures with which they are used.

Competition between the need to survive and the need to conserve the resource
In the past, fisheries management has been much preoccupied with notions of 'sustainability', envisioning sustained fisheries that can continue regenerating themselves indefinitely. Attempts to bring this perspective into the fold of social science have emphasised the need to ensure that resources are not only available for present generations, but also future ones (cf. Redclift, 1984, 1987; WCED, 1987). Since these early discussions, much consideration of this problem has focused on the livelihood as the unit of analysis around which sustainability must turn (Chambers, 1983; Chambers and Conway, 1992; Conway and Barbier, 1990). A livelihood is:

> '...a means of securing a living...Encompassed in a livelihood is the totality of resources, activities and products which go to securing a living. It relies on ownership of, or access to products or income-generating activities. A livelihood is measurable in terms of both stocks – that is the reserves and assets – and the flows of food and cash' (Conway and Barbier, 1990: 117).

The notion of *access* to resources is the most important factor in determining whether or not the pursuit of a livelihood is successful (cf. Ellis, 2000). It follows, then, that where a resource on which a community relies becomes scarce, then access to it is curtailed. As resources become ever more scarce, then the measures that people will adopt to try and procure resources may become increasingly more desperate, and more difficult for the resource concerned to sustain. This trend may then reach a point where the pursuit of livelihoods actually starts to undermine the ability of a resource to regenerate itself.

If Medard and Geheb's (2000) assessment is followed, the fisheries management objectives of the Beach Management Units (BMUs) are actually beside the point insofar as fishing communities' pursuit of livelihood goals are concerned. What is the point is that fishing communities will evaluate BMUs in terms of their threat to, or support of, livelihood claims. If, for example, one such claim is easy access to juvenile fish (as would appear to be the case; see Geheb *et al.*, 2000c), then fisheries management objectives are violated. In addition, if this is a livelihood claim of Tanzanian fishing communities then whatever BMU strategies exist to impede the claim will be by-passed.

The above message – that the pursuit of livelihood claims may not necessarily translate into a sustainable fisheries resource – applies to all fishing communities around the lake. As Baland and Platteau (1996) argue, it does not follow that resources left in the hands of small-scale communities of resource users will automatically be optimally managed. In its present context, this would seem true for Lake Victoria's fisheries, but does not preclude the possibility for fishing community participation in optimal management were the lake's regulatory context to be altered.

An overemphasis on managing the biological basis of the resource

Traditionally, fisheries management has fallen within the purview of the biological and limnological sciences. Most of the management models derived from these disciplines require a substantial data input (such as stock assessment or frame survey data) so as to inform management how best to allocate fish stocks between users and the supplies needed for stock regeneration. The data demands of these models, however, are expensive and require highly trained personnel. There is no guarantee, therefore, that such data collection activities will occur as management demands them in the adverse economic, social, political and cultural environments of Africa. In any case, there is considerable doubt that management based on these strategies actually works (cf. Crean and Symes, 1996; Ludwig *et al.* 1993). It makes sense, therefore, to introduce the idea of 'appropriate imprecision'. The dogged pursuit of 'measurable' data on fish stock dynamics is both a distraction and an excuse for management systems unable to deliver regulatory outputs. The fact that several hundred fishermen spread around a fishery tell management that catches have declined both in terms of individual fish size and volume, may be imprecise, but is little different from a stock assessment survey delivering the same message. One important difference is that collecting the data from fishermen requires a lesser commitment of human – and possibly financial – resources.

Management may also raise concerns about species loss and reduced bio-diversity (Harris, 1998). In this aspect Lake Victoria is no exception and there are those that contend that the study of evolution has been irrevocably damaged by the crash of the haplochromine species flock in the lake (Witte *et al.*, 1994; Goldschmidt *et al.*, 1996). However, the principal protagonist in its demise, the Nile perch, is also the backbone of a spectacular fishery, employing many thousands of people and the basis for a multi-million dollar export industry. As Aarninck (1999) comments, in the cash- and staff-strapped fisheries departments of Africa, it makes sense to prioritise management concerns, and a guide for such an activity should be based on what is feasible and affordable. Suggestions to, for example, restore the haplochromine species flock to its former, pre-Nile perch, condition fare poorly against such a guide. Stock assessments and frame surveys should be judged similarly.

'Resource problems are not really environmental problems...[but]...[h]uman problems that we have created at many times and in many places, under a variety of political, social and economic systems' (Ludwig *et al.* 1993: 36). Biological manifestations of over-exploitation are symptomatic of wider economic, social, political and cultural events and processes. Access to resources and the ways in which they are used typically have more to do with the relationships between members of the resource using community, and their relationships with the outside world, than they have to do with the community's relationship with the resource *per se.* As in medicine, it makes more sense to treat the causes than it does the symptoms, and it is therefore difficult to explain why it is that fisheries management, both within East Africa and elsewhere, is so preoccupied with biological inputs. Experiences associated with the collapsed North-West Atlantic and threatened North-East Atlantic cod stocks, all of which were heavily managed, indicate the non-viability of managing fisheries solely on biological principles. Fisheries management revolves around the regulation of resource exploiters and their technology, and not about managing fish stocks.

The fisheries research institutes of East Africa are dominated by the physical sciences. In Kenya, there are four social scientists of researcher rank and above on the staff of the Kenya Marine and Fisheries Research Institute's (KMFRI) station in Kisumu; in Tanzania there are two, and in Uganda one. Nevertheless, research priorities amongst the institutes continue to call for a biological research bias. As one former KMFRI director has commented, '[f]uture research on Lake Victoria should emphasise on [*sic.*] fisheries biology and environmental studies...to guide policy makers, investors and managers on matters of fisheries management and conservation' (Okemwa, 1995: 190).

It is not the intention of this chapter to argue that biological studies in fisheries management are unimportant. What it is concerned with, however, are management systems *driven* by biological information inputs. In a region with limited expertise and funds, the development of management systems that rely on stock assessment data is problematic. In any case, the cost of such exercises must be evaluated against the fact that such data reveal only trends in indices of stock size and composition, and not the reasons for stock size and composition change. Nor does such information equip managers with the tools necessary to tackle the problems that cause them.

A related problem lies in the relationship between fisheries research institutes in the region and fisheries departments. In Uganda, as part of the country's decentralization policies, district fisheries personnel answer to district offices, and not to DFR headquarters. The latter has a far greater regulatory role, while district personnel are more occupied with collecting various duties and levies. As in Uganda, Tanzania's Fisheries Department has both regulatory and revenue collection roles. Insofar as the latter is concerned, the Department's activities fall under the purview of the Ministry of Tourism and Natural Resources. As far as revenue collection is concerned, however, the Fisheries Department's activities fall under the scrutiny of the Ministry of Local Government. In Kenya, the research institute and Fisheries Department fall under the same ministry, but both have the right to carry out research. The same is true in Uganda, where both the Department for Fisheries Resources and the Fisheries Resources Research Institute have the right to carry out fisheries research. Fisheries research on Lake Victoria should not be curbed and limited to a certain set of institutions, but research and enforcement need to be separated, and research should be determined by the demands of managers.

Underestimation of community capabilities with respect to their role in the management process

In many respects, fishermen and their communities are seen as ignorant, slovenly and untruthful. This results in two managerial difficulties. On the one hand, fishermen and their communities become criminalised, both in the minds of administrators as well as in their own. Fishing communities come to perceive that there is little that they can do that is right, and understand fully that this is what the state understands of them. In such circumstances, relationships between fishing communities and the state are not as good as they could be.

On the other hand, the difficulty leads to government officials regarding education as a panacea for the management problems of Lake Victoria. Fishermen must be educated so that they know what they are doing is wrong. In all probability, however, fishermen are fully aware of what they are doing. They are also aware of trends within the fishery and fully able to advise fisheries professionals about these (cf. SEDAWOG, 2000; Geheb, 1997). If education is not at the crux of Lake Victoria's problems, the late's common property status presents severe managerial difficulties, ensuring that fishermen know well that whatever fish they leave behind on one day will only be caught by someone else the next day.

These problems of attitude are not restricted to the fisheries administration alone. Amongst Kenyan fishing communities, the pervasive view is that the government lies at the centre of management, which precludes the inclusion of communities in the management structure (Lwenya, Ouko and Onyango, pers. comm.).

The capacity to deliver an effective regulatory service

'Capacity', here, is considered in terms of the ability of a fisheries department to deliver a service, in this case, fisheries regulation. In some cases, this ability is limited by funding difficulties. The Kenyan Fisheries Department earns some US$ 70,000 from the sales of fishing licences and boat registration (Government of the Republic of Kenya *et al.* 1995). Despite additional contributions from central government and, occasionally, international development agencies, the total cost of wages for all of the Kenyan Fisheries Department personnel stationed on Lake Victoria comes to a yearly average of some US$ 413,000 (Government of the Republic of Kenya *et al.* 1995). As central government finds itself increasingly unable to meet budgetary demands, Fisheries Department personnel only draw wages, and have not the funding to provide services. Indeed, recurrent operating expenses within the Kenyan Fisheries Department represent just 9% of its total budget, while the remaining 91% is used exclusively to pay wages (Government of the Republic of Kenya *et al.* 1995).

Staffing constraints are also concerns, although not in Kenya where the Fisheries Department has 611 staff around its portion of Lake Victoria (Government of the Republic of Kenya *et al.* 1995). In Uganda, however, the Fisheries Regulations and Control Unit (FRCU) has to police 43,941 km^2 of water with 35 staff (Kiiza, 1998).

Irrespective of staff numbers, the assumption is often made that regulations will be effectively and unambiguously upheld. In Cleaver's (2000) important work on the management of water resources by a community in Northern Zimbabwe, she demonstrates how decisions are always arrived at via a lengthy process of negotiation between the parties involved. Similar negotiations occur all along Lake Victoria's shores. A problem occurs, however, when the regulations of the state are subjected to similar processes. Olivier de Sardan (2000) argues that this is certainly the case in many African government departments. In many respects, such negotiated outcomes may seem sensible. As one senior Kenyan Fisheries Officer comments, 'The Fisheries Department is aware of the difficulties of getting work elsewhere, and needs to allow for some opportunities to be open for [fishermen]. It was hard for a Fisheries Officer to prosecute offending fishermen...we are aware that by doing so is as good as denying the fisherman's family an income...' (quoted in Geheb, 1997: 81). This difficulty, of course, harks back to the problems inherent in enforcing regulations, which might curb a community's pursuit of its livelihood claims.

The negotiated outcome to the region's fisheries regulations does not, however, end with pity. East Africa's British colonisers were well aware of the shortcomings of sending staff from one ethnic group to police state regulations in their home areas, and contemporary fisheries officials are cognisant of the difficulties surrounding the 'nepotistic pardon'. Being related to the official concerned, furthermore, may not only improve the chances of being pardoned, but also ensure that the bribe negotiated is not too high. At the most pessimistic level, the regulations in place on Lake Victoria have introduced new money-making opportunities for under-paid fisheries staff, while at its most optimistic, they have reduced enforcement to an apathetic level. This apathy pervades the fisheries regulatory structure. It is tempting to suggest that while there are problems in enforcing regulations that user communities have trouble believing in, it is even more difficult when enforcement staff themselves do not believe in them as tools in the fight for a sustainable fishery.

These problems are not only restricted to fish landing sites. Even components of the fishery that are relatively easier to regulate, such as the fish processing factories, would appear to be neglected. At virtually every fish landing site visited by the teams working under the LVFRP project, fishermen actively sold fish of 2 kg and below to Nile perch fish processors, and even claimed that the factories specifically sought fish in this size range. Nile perch of 2 kg weight have typically not yet bred. Mitullah (1998) argues that the reliance of fishermen and the fish processing factories in Kenya on under-size Nile perch is dependent on the 'partisan' position of Fisheries Scouts. Scouts own boats and illegal nets and themselves catch and sell under-size Nile perch. 'The fishermen normally organise a small delegation with "*chai*" [a bribe] to talk to fisheries officials during closed season'" (Mitullah, 1998: 56).

Wilson and Medard (n.d.: 76), commenting on the relationship between fish processing factories and the Tanzanian Fisheries Department have the following to say:

> 'There is extensive distrust between the government agency personnel and the processing plant executives. This has ideological origins stemming from Tanzania's history. It is also a result of corruption that both groups participate in; and that some factory managers see as almost the exclusive determinant of government decisions."

With fish processing factories absorbing as much as 80% cent of Tanzanian Nile perch catches (Gibbon, 1997), the illegality of their operations is deeply worrying. Official discussions regarding the problems of regulation on Lake Victoria often focus on the idea of 'harmonisation': that if all of the countries sharing Lake Victoria have the same regulations, then somehow the regulatory problems of the lake will be solved. In much the same way, there exists within the region a pervasive belief that the enactment of laws and regulations at the parliamentary level will automatically translate into obedience and compliance on the ground, with no intervening act of enforcement in between.

Partly as a result of these problems, in June 1994, the Lake Victoria Fisheries Organization (LVFO) was created. In December, 1996, the LVFO was brought under the control of the newly (re-) formed East African Community (EAC). The potential for the LVFO to play a vital role in the co-ordination of research and regulation on Lake Victoria has, however, not been realised. The EAC exists in little more than name at present, and it is from its headquarters in Arusha, Tanzania, that the LVFO draws its powers. The region's fisheries departments and institutes are under no obligation to answer to the Organisation, let alone answer its calls.

Unrestricted access to a common fisheries resource

All the resources of the lake are subject to minimal control on who might access them and where this access might take place. Even at a community level there are strong perceptions that the fisheries resources are for all. In practice this is demonstrated by the equanimity with which resident fishers greet the arrival of 'outsiders'. Thus, fishers may avail themselves of the facilities and order of a new landing area provided they make the necessary representations to the local authorities and bring with them a letter of introduction from where they last operated as fishers.

The view of the lake as there for all is also reinforced by the general antipathy to the creation of boundaries to restrict access or separate different fishing activities. We know from other fisheries systems that, as effort increases and problems develop, recourse to boundaries to limit access become paramount amongst the tools used by fishery managers. Yet the fishers of Lake Victoria do not seem ready for this, still hoping that other solutions might be found.

As if the access and ownership problem of the lake's fisheries were not enough, this situation must be set in the reality of a number of different sectors, and many diverse user groups, accessing all the resources of the lake. Hence, the water in which a fish stock is contained is also used for drinking water, waste disposal, watering livestock, bathing etc. A management plan, in seeking to protect a fish stock, must not violate the right of users to exploit the resource for purposes other than fishing (cf. Swallow *et al.* 1997).

The building blocks for management

In summary, we have argued that the three main problems with the management of Lake Victoria's fisheries are: (a) the failure of Ugandan fishing communities to grasp the opportunities presented by their Government in the form of the 1997 Local Governments Act; (b) the creation of BMUs in Tanzania which have no rights to contribute to the design of regulations, weakening the potential efficacy of the BMUs; and (c) the failure to incorporate communities into the management structure of Kenya's Lake Victoria fisheries. We have argued that the following are additional and/or complicating factors: regulatory ambiguity, a reluctance by the state to transfer power to fishing communities, a reluctance by states to endorse local-level regulatory institutions, a contradiction arising between the pursuit of livelihood goals and fisheries management objectives, an excessive emphasis on biology (and stock assessment) in the fishery's management and research structures, an underestimation of the knowledge and abilities of fishing communities, the minimal capacity of formal fisheries institutions within the region to regulate the fishery and problems of ownership/access to the resource.

We contend that it makes no sense to consider calls for greater enforcement, or the enactment of more stringent rules (cf. Dunn and Ssentongo, 1991) on Lake Victoria because there is no evidence that these interventions have worked before. If the pursuit of livelihoods is viewed, in part, as a series of informal institutions aimed at maintaining and improving access to life-sustaining resources, then there may well exist a lack of 'fit' between formal and informal institutions on Lake Victoria. Additional regulations of these varieties will only serve to exacerbate this mismatch and, at worst, actually undermine the ability of fishing communities to obtain livelihoods. As such, it makes sense to consider radical alternatives.

We propose, therefore, a management plan comprising two broad characteristics: the first is that it should not undermine the efficacy of the livelihood decisions of fishing communities. The second is that it should contain adequate opportunities for managerial concerns to be vented and scrutinised, along with structures to ensure that such scrutiny actually influences managerial decisions. The plan should be able to provide solutions to the various dilemmas discussed above. With this in mind, we suggest that the management plan must therefore contain the following, critical, components:

(a) At its core, the plan should be founded on beach institutions, which will be at the front line of regulatory enforcement.

(b) The influence of the state, fish processing factories and other stakeholders should be felt at all levels of the plan's administration in a negotiating – rather than a voting – capacity.

(c) Laws and regulations under the plan should, as far as is possible, be generated by negotiation and consensus. At its base, the plan should be minimally prescriptive so as to ensure that communities of resource users have the greatest possible opportunity to develop regulations that suit their own locations, culture, conditions and resource dynamics. As meso- and regional levels occur, prescriptions increase to reflect greater levels of government and private sector involvement. These prescriptions include suggestions for voting and delegate selection.

(d) The plan must in no way impinge on the rights of other water users.

(e) Beach institutions should be 'nested' within a wider framework providing, in the main, support and facilitating services. It must also be able to feed into other political and administrative structures not necessarily related to fisheries.

(f) The plan must be able to survive with minimal scientific and financial inputs.

(g) The plan must be adequately flexible and amenable to change so that it can cope with fluctuating economic, social, political, ecological and limnological conditions.

(h) The plan should be amenable to gradual implementation, so that dominant political sensibilities are not offended, and communities may become used to their new responsibilities.

(i) The plan must be backed by a comprehensive legal package guaranteeing communities the right to be involved in fisheries management. The package should also establish and reinforce the independence of the plan's components from other political processes and interests that might seek to undermine the plan's efficacy.

(j) The plan must contain within it the legal obligation of all actors to be transparent, and that all actors must design ways to assure that this is achieved.

(k) The plan must contain within it the promise of dissemination, so that all actors are informed.

In order to administer the plan, we designed a three-tiered hierarchy for fisheries administration which contains within it the right to seek horizontal support and influence (Figure 20). We have selected the three tiers as ones that lie at the community level, at the meso-level (district) and the regional level. In some respects, there may be some benefits in having additional, intermediate, levels (Asila, pers. comm.), but this will serve to increase the costs of administration.

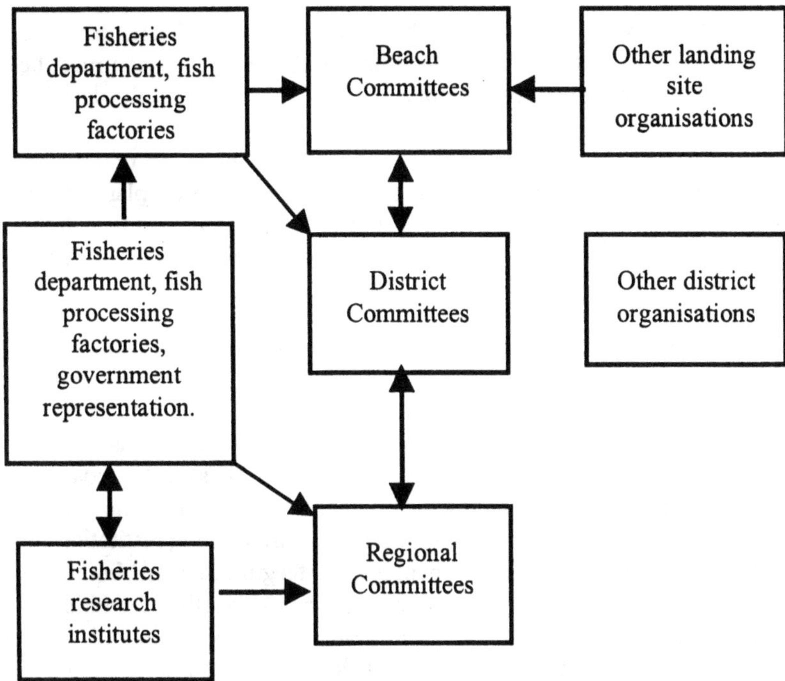

Figure 20: Overview of suggested fisheries management plan institutions and administration.

While each of the tiers has obvious administrative functions, their primary roles are as foci for discussion and negotiation. Experience gained under the Lake Victoria Fisheries Research Project (LVFRP) suggests that no matter what the level of administration, outputs are the result of (often protracted) negotiation, and it therefore makes sense to propose an administrative structure that maximises the opportunities available for this. In addition, having such forums for the discussion, enactment, implementation and enforcement of fisheries regulations is crucial if the stipulation that face-to-face communication between actors is to be met. Ostrom (1997) argues that such communication is a vital component in the generation (and increase) of trust between players (Ostrom, 1997).

An additional problem that the plan attempts to address is the manner in which decisions are arrived at. At formalised levels of administration, the use of a majority vote may well be acceptable to participants, but at community, informal, levels, only outright consensus may be considered acceptable (Cleaver, 2000). It is therefore necessary that the management plan enables either method to be useful, and should therefore only stipulate that (a) decisions are necessary; (b) that the manner in which these are agreed upon must be determined by the representative group concerned; and (c) that the decision must be conveyed.

In the sections that follow, we examine each of the three tiers of the plan, and argue that to greater and lesser degrees, each tier enables the plan to fulfil the background conditions listed above, and consider the ways in which the plan deals with, or ameliorates, the dilemmas discussed in the first part of this paper.

The role of Beach Committees in developing and implementing the management process

All landing sites on Lake Victoria have some kind of beach committee, be they those already present in Kenya, nascent Beach Management Units (BMUs) in Tanzania, or Landing Management Committees (LMCs) in Uganda. We propose that such Beach Committees (BCs) should form the backbone of Lake Victoria's fisheries management. The use of these pre-existing institutions in the plan is very important because not only does it legitimise them, but it also ensures that already existing institutions are maximally utilised within the management framework. Communities already have experience with these institutions, and while an augmentation of their responsibilities may prove controversial, their presence at landing sites no longer is. The responsibilities and other functions of the BCs are summarised in Figure 21 below.

It must be reiterated, however, that the BCs must be as representative as possible. Without wishing to suggest how the committees are selected, it is important that all members of a landing community contribute to the selection process. This, in turn, necessitates that a community defines itself. Provided communities can do this, there seems little reason why they cannot agree on the overall structure of the BCs, including the number of members who should sit on it, what their individual roles should be and how the BC itself will arrive at decisions that have the community's blessing.

BCs should meet regularly so that the number of opportunities available for the expression of grievances is maximised. Through such meetings, communities themselves should decide what regulations they wish to implement, monitor and enforce. Fisheries department extension personnel must contribute to negotiations, making known scientific consensus regarding the fishery, and recommendations regarding the management of the fishery. It is also important that fish processing factories (FPFs) are present to try and influence regulation and/or prices. Other stakeholders can also be present as the BC sees fit. Whether or not these various interest groups should have decision-making powers on the BC is debatable, and is possibly best left to the communities to decide.

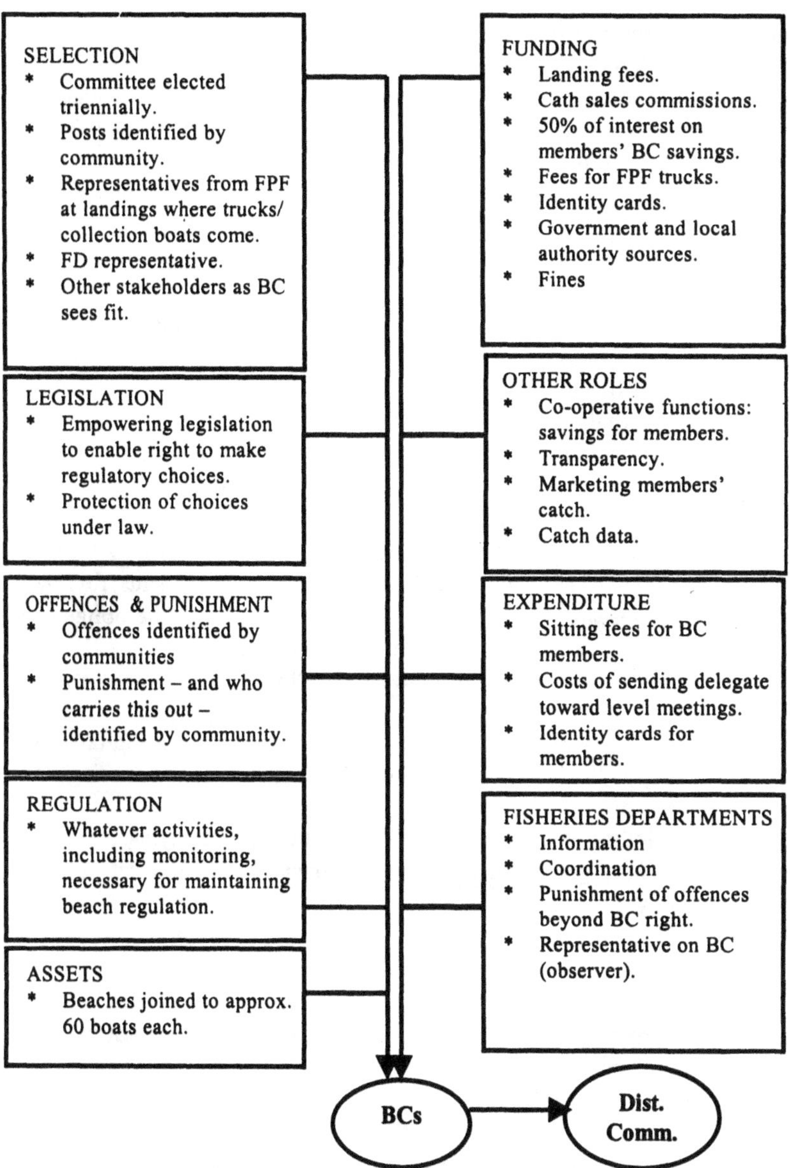

Figure 21: Components of Beach Committee organisation, roles, jurisdiction, fund-raising and expenditure.

The point with these meetings is that the outcome of these negotiations should be one that in some measures meets all the desires and demands of all the stakeholders represented on the BC. An additional potential benefit of including a diversity of stakeholder representatives on the BC is that they may monitor each other for defaulting on BC agreements.

The rules and regulations agreed upon by the BC should be tailored to suit the ability of the BC to implement and enforce them. Those regulations that fall beyond this threshold should be past on for administration by District or Regional Committees.

We propose that BCs are (s)elected triennially. This duration should be sufficiently long that the incumbent BC will have been able to carry out its work, but also sufficiently short that it is subject to frequent electoral scrutiny.

Management of the fisheries requires reliable catch/effort information. An improved relationship between BCs and research institute personnel is important in this area. Ideally, the research institute would negotiate with communities what data they will collect.

Communities, through their BCs, should also decide what they wish to know about, and pass these messages on to the fisheries departments, for relay to the fisheries research institutes. The latter should, by law, be obliged to research those issues identified by fisheries departments.

The authors believe that the structure and outlook of BCs should enable the following activities:

(a) the identification of regulations which they believe are just and fair, and which they believe they are capable of implementing, monitoring and enforcing.

(b) The sanctioning of these rules in ways that the community agrees upon, and which are graduated in a manner considered appropriate by the community involved.

(c) The provision of representatives to District Committees, who could be individually selected by the Beach Committee.

How BCs punish offenders should be determined by the communities themselves. It may, however, be the case that communities do not wish to have this role, or that they feel that they should not adjudicate certain crimes. In these cases, BCs should determine who should punish offenders, and/or at what point crimes become so serious that they must be referred elsewhere. Drawing the BCs more strongly into the control, a monitoring and surveillance process is an essential pre requisite of a realistic management approach. Referred cases can be passed on to the police or, in Tanzania, *sungusungu* vigilante groups. Alternatively, they may be passed on to District Committees for consideration. This particular area of administration could well be one in which communities assume gradually augmented responsibilities.

The funding of these committees should be derived from beach contributions in the main. There are very few fishermen's co-operatives in Uganda and Tanzania, but these are common in Kenya. Having a co-operative at a landing area represents an alteration of local power structures, which can, in some cases, lead to local level confrontations when two or more power structures exist (cf. Geheb *et al.*, 2000a). We recommend, therefore, that BCs assume co-operative responsibilities, including the collection of commissions on catches sold at the landings, and the provision of savings accounts to members. As a fee, the BC might levy an account holding fee on interest paid on accounts claiming, say, 50% of interest. Additional funding sources can be levies applied to trucks from the fish processing factories coming to the beach, and similar levies on migrant fishermen. Fines imposed on offenders are an additional source of revenue. The BCs may also decide to seek additional sources of funding such as, for example, donors, fish processing factories or local authorities, but the functioning of the BCs must at no time become contingent on these sources.

A vital component of this process is that it be transparent. With large amounts of money passing through, and being administered, by BCs, it is crucial that they should not make the same mistakes that have afflicted many Kenyan fishermen's co-operative societies. Geheb (1997) argues that the success of certain Kenyan co-operatives is contingent on transparency between management and members, and that one way of providing this was through the display, on a regular (weekly) basis, of the status of accounts and other assets. In addition, the details of monies held on behalf of members could be displayed by tacking these details on the co-operative office wall for all to see. A similar method could be used (or even prescribed) by the BC. Failure to meet these criteria should be sufficient grounds for the dissolution of the BC. Our reasons for suggesting this is that we believe that it is necessary – through transparency and frequent electoral review – that BCs be continuously assessed by their communities.

There are an average of 28 boats per landing site on Lake Victoria (calculated from FSTC, 2001). We suggest (arbitrarily) that the jurisdiction of the BCs should extend to around 60 resident boats. Where a landing has less than 60 boats, it should merge its organisation with additional landings until the total is met. We suggest that the main unit of organisation in the management plan should be the landing site, and because reciprocity amongst groups of resource users tends to improve when the numbers involved are fairly small, then it makes sense to try and ensure that the populations of administrative units are kept as small as possible. In addition, we believe that the transfer of ownership of the landing to the BC would be of benefit both from a regulatory perspective, but also as a source of collateral should the BC wish to negotiate for loans.

Research carried out by the Lake Victoria Fisheries Research Project (LVFRP) and others (cf. Geheb, 1997) suggests that fishing communities believe that it is very important that an adequate basis for the identification of fishers – especially migrant fishers – exists. In particular, fishing communities believe that such a means of identification is important to ensure that those outsiders welcomed to their landing sites are not thieves and will not steal their gear. At present, this facility is catered for mainly by letters of introduction, which migrant fishers present to landing authorities when they arrive. These will identify the name of the fisher, the number of gear and crew that he brings with him, and that he left his previous beach with no ill feeling and a 'clean record'. In some areas in Uganda, landings have started issuing fishermen with identity cards, which display their photographs. This is often considered a normal part of the beach committees' portfolio of activities. The importance of these cards is that fishermen can travel freely between landings as they pursue better catches. If they at any point steal, or violate landing rules, the cards can be withdrawn and the offender prevented from migrating any further.

In view of the demand for such identification, and a means for verifying the trustworthiness of migrants, along with the need for a means with which to identify community members, encourage a sense of 'belonging' to communities, and a possible basis for future effort regulation, we suggest that identity cards for fishing community members are a viable way of meeting these needs. We propose that these cards be issued to all members of communities, and could fulfil the following functions:

(a) only those fishers in possession of a card will be allowed to fish, or sell their fish, from landings at which they are not resident.
(b) BCs could retain the right to revoke cards in the event that the holder commits an offence, hence impeding further migration and involvement in fishing activities.
(c) BCs could retain the right to control the number of cards issued.
(d) A migrant fisher travelling to another landing site is subject to the regulations of that site.
(e) The cards could replace current licensing and boat registration regulations.
(f) They could serve as a basis for future effort controls.
(g) The BCs could levy small charges for issuing the cards, which could serve as income.
(h) The cards could permit holders to participate in BC selection events.

The community as a whole, which will have a direct input in budgetary matters, should decide on how the BC's funds are spent. For the sake of transparency, communities should revisit budgets frequently – possibly every six months. The BCs therefore should consider the following potential costs:

(a) Regulatory and enforcement costs.
(b) The costs of creating a beach office, stationery and other office requirements.
(c) The payment of the chairperson and one other to attend District Committee meetings.
(d) The payment of district and municipal taxes and levies. Communities may also wish to consider remuneration for BC staff.
(e) All surpluses must be declared and served as dividends to card holding members. The research experience of the LVFRP has been that the realisation of rapid cash benefits may play an important role in generating confidence in management systems.

The relationship between the BC and the higher levels of governance will be critical to the success of the participatory management strategy. The 'nesting' of local level institutions within wider administrative and resource monitoring structures is often perceived as vital for the successful functioning of local level management structures (cf. Ostrom, 1990). Structural features to facilitate this should include:

(a) Once every six months, the District Committee will meet, and the BC must send its chairperson.
(b) An additional individual, nominated by the community, should also be sent to ensure that the chairperson properly represents community interests. His/her discussions at the district committee are considered below.
(c) The BC must be able to fund these delegates.

The role of District Committees in developing and implementing the management process

The district is a convenient administrative unit around which to organise a meso-level forum for fisheries management. There are 31 districts along Lake Victoria's shore, and 1,493 landing sites (FSTC, 2001). There are, therefore, an average of 48 landing sites per district, and District Committees would then comprise approximately 96 members (two per BC) each. At this level, other stakeholders should also have representation, and, as a guide, these could include fish processing factory (FPF) representatives and fisheries department personnel. Again, the role of these other stakeholder groups on the Committee would be to attempt to influence it in particular directions. If the district concerned also contains lake-side municipalities, these should also be represented. The District Committee should be empowered to decide whether or not other groups should be represented upon it. At this level, it is not inappropriate that decisions are arrived at by vote, which could be carried out by secret ballot, overseen by the district administration. Figure 22 provides a summary of components of District Committee organisation, roles, jurisdiction, fund-raising and expenditure.

The *raison d'etre* of the District Committee is to pass regulations concerning fisheries at the district level, in particular over issues such as fish passing between or through districts by road, or regarding migrant fishermen entering the district. The District Committee should also have the power to pass regulations to resolve inter-community and district conflicts. The District Committees should consider whether they are capable of taking on the following roles:

(a) To meet and consider district-wide trends in their fisheries and other related problems.
(b) To consolidate all scientific data collected.
(c) Offences that the BCs have felt are too grave for their consideration will be dealt with by the District Committee or referred to the Regional Committee.
(d) Receive advice from an attendant Fisheries Department officer, who may also use this opportunity to brief members of district-level changes to the fishery.
(e) To listen to requests and receive advice from FPF personnel.
(f) To consider all matters on an inter-district nature – such as, for example, fishers from one landing breaking the rules of another – that lie beyond the jurisdiction of BCs acting alone.
(g) The District Committee should not have the power to over rule or over turn BC regulations, but be in a position to counsel and advise communities on the regulations they suggest.
(h) To seek external funds from government and/or other organisations if the Committee sees fit to do so.
(i) To select a person from each of the stakeholder groups represented on the Committee to send to an annual Lake Victoria regional fisheries management meeting convened by the Lake Victoria Fisheries Organization (LVFO).
(j) Each meeting will generate a report to be forwarded to the LVFO. The report will contain a summary of scientific data collected; a list of all BC regulations passed within the district; a list of all offences that have occurred within the district and how these were sanctioned; and a series of recommendations concerning any or all of the information presented. This report shall be passed by a majority committee vote.

The Regional Committee

This will meet once annually under the auspices of the LVFO, which will rotate the meeting place amongst the three main cities bordering the lake (Kampala, Mwanza and Kisumu) and its headquarters in Jinja.

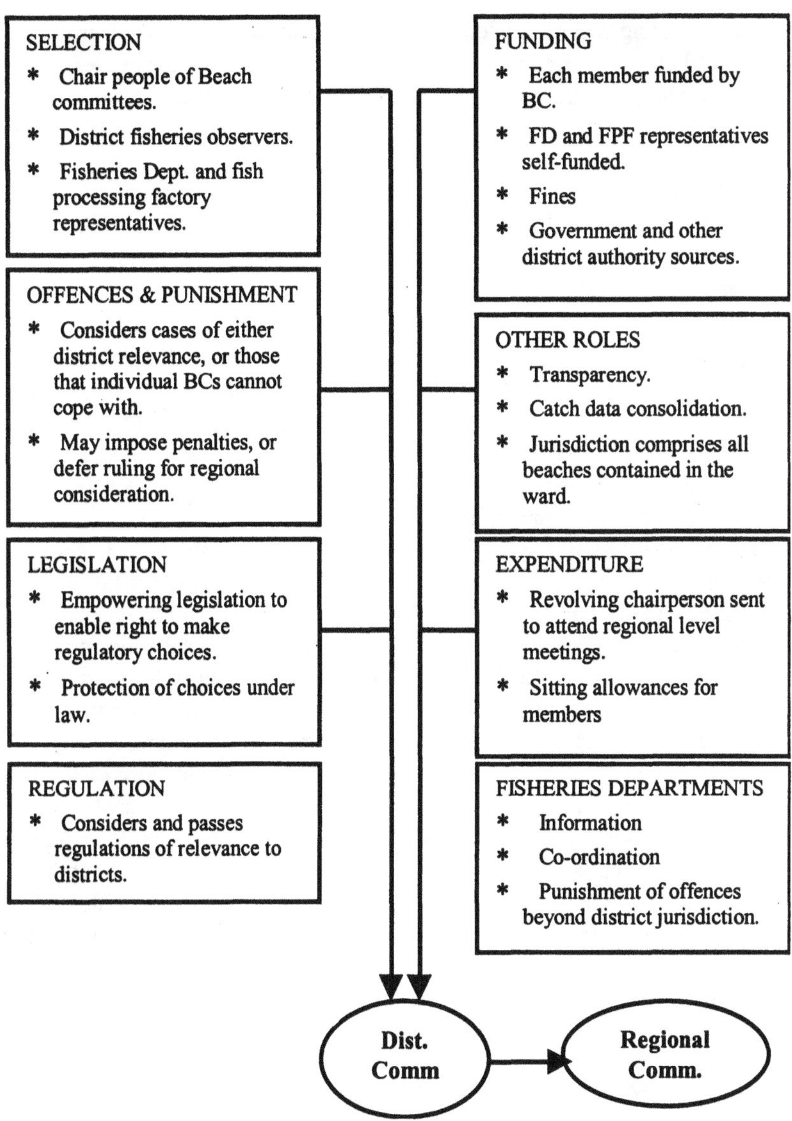

Figure 22: Components of District Committee organisation, roles, jurisdiction, fund-raising and expenditure.

Delegates meeting will be funded by the LVFO, and will be drawn from each of the District Committees, selected as described above. Delegates shall also include representatives from fisheries research institutes, and from other sources (such as ministers or other high-level administration functionaries). Components of regional committee organisation, roles, jurisdiction, fund-raising and expenditure are summarised in Figure 23.

It is these delegates (fishing community representatives, Fisheries Department personnel, fish processing factory representatives, municipality representatives and fisheries research institute personnel) that shall form the electorate of the Regional Committee. The number of participants shall be limited. In the case of Fisheries Department and Beach Committee representatives, upper limits should be tagged to the total number of districts on the lake shore. Municipal representatives should be limited to the total number of municipalities on the lake shore. Fish processing factories can send no more voting members than their total number, and fisheries research institutes may send only staff of senior researcher rank and above.

The principal task of the Regional Committee is to consider the fisheries matters of Lake Victoria as a whole. Its main legislative powers lie in the realm of, for example, pollution at the national and regional levels, upstream consequences of their actions, such as those concerning the Nile, or the inflowing Kagera River. Within this remit may also be fisheries legislation of a regional nature. This latter caveat, however, must be carefully constructed so that it does not undermine the validity and strength of community-level regulations. The jurisdiction of the Regional Committee should also concern itself with matters of quality assurance, the Nile perch export industry and other areas of marketing. Other roles of the Regional Committee could include reviewing all data collated by the LVFO from the District Committee reports that have been received. The meeting will enable candidates to consider and debate the findings, and to vote on whether or not the final report should be issued. Insofar as fisheries legislation is concerned, the Regional Committee can only make recommendations that the District and Beach Committees will consider whether or not to follow.

Constraints to implementation

There exist within the Lake Victoria region tremendous concerns over granting communities powers of regulation, enforcement and design. If it is assumed that the objectives of fisheries management on Lake Victoria are related to the sustainability of the resource, then such a transfer of responsibilities from the state to fishing communities can, at its worst, merely replicate current levels of apathy.

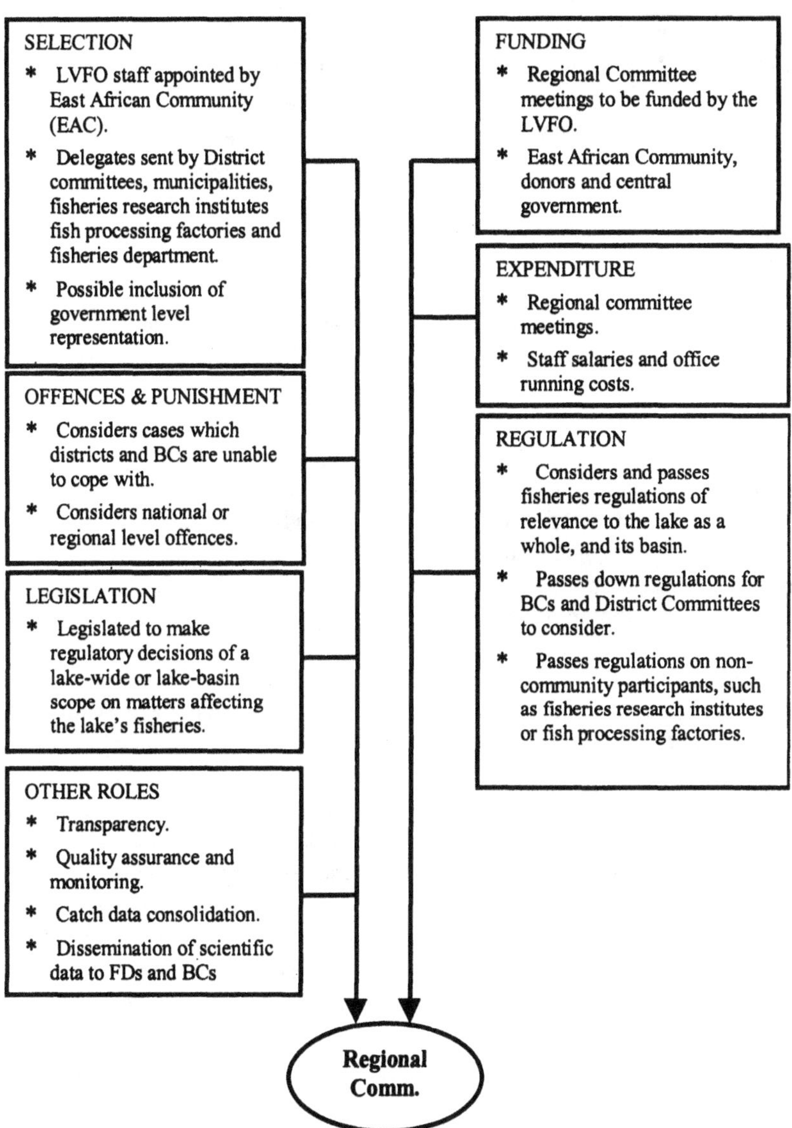

Figure 23: Components of Regional Committee organisation, roles, jurisdiction, fund-raising and expenditure.

If, however, the objective of fisheries management is to retain state discretionary rights over fishing communities, then it makes sense to consider possible ways in which changes to the quality of this power could be ameliorated. Perhaps the best way of ensuring that honour is met in all respects is if the plan is implemented via a designed step-by-step strategy, gradually augmenting the responsibilities of communities over time, and gradually changing the quality of discretionary power. Such an implementation strategy can, in addition, be pegged to performance. Hence, if communities fail to identify and apply regulations, the plan can then be paused, and problems addressed.

An additional way in which the augmentation of community-level responsibilities can be introduced is to structure implementation around a single regulatory concern which communities are prepared to act upon.

Throughout the work of the LVFRP, the theft of gear was repeatedly indicated as amongst the most serious of fishing communities' concerns. By granting communities the right to guard their nets, to arrest and punish thieves, the necessary groundwork for assuming responsibilities may be laid, and can be built upon as the plan matures.

Implementation must also be pegged to information provision – ensuring that all fishing communities are aware of their rights, the objectives and hopes of the plan and the directions in which it may head.

Conclusions

This chapter has argued that problems in the management of Lake Victoria's fisheries are primarily derived from the absence of co-management between the state, fishing communities and industrial fish processors. The paper suggests that these problems manifest themselves in each of the lake's bordering states as 'meta-problems' from which all further regulatory problems emanate. In Uganda, these problems relate to the ignorance of fishing communities of their rights under the law. In Kenya, these relate to the absence of any degree of community participation in the administrative structure; and in Tanzania, they relate to the provision of enforcement powers to communities without including them in the design of the regulations they are supposed to enforce. We argued that these problems have led to: related difficulties of regulatory ambiguity; state-community dichotomies; the failure to legally endorse local-level institutions; the gap between the fulfilment of livelihood objectives and fisheries management objectives; an over-emphasis on the biological sciences in fisheries; an underestimation on the abilities of fishing communities; the under-capacity of state regulatory organisations; and finally the open access/multiple user aspect of the lake.

The sections that followed described a possible management plan that could ameliorate these difficulties. We do not pretend that the plan is comprehensive, but suggest that what is presented may form the foundation of such a plan. We propose a three-tiered administrative system that has at its core Beach Committees. These are selected in ways that the communities involved see fit, and should, we believe, have upon them representatives of the fishing communities, fish processing factories, fisheries departments and any other groups that the fishing communities feel should be represented. We believe that the decisions of the BCs should be arrived at through means determined by the communities involved, and that the voting powers of the representatives should also be determined at these levels.

The BCs should be at the forefront of fisheries administration on Lake Victoria, and their decisions will inform those of the next tier, the District Committees, whose roles are to make recommendations on fisheries matters of district-level concern, and to provide an over-arching service to the BCs, informing them of district-level fisheries trends, and events of concern to landing sites. The District Committees also have the task of deciding upon cases that the BCs feel that they cannot themselves handle. Finally, these Committees act as a half-way house between the fishing communities and a Regional Committee, to which they will select and send representatives.

The Regional Committee will meet under the auspices of the Lake Victoria Fisheries Organization, and its responsibilities are similar to those of the District Committees, except that its brief shall extend to national and regional levels.

The logic behind the above design, and our suggestions for membership on the three committees, is derived from our belief that regulatory outcomes on Lake Victoria should be the product of a negotiated process between various stakeholder groups. It is only if such a process occurs that we can expect fishing communities to view fisheries regulations as valid and acceptable, and hence, worth enforcing.

Acknowledgements

This chapter represents an output from the Lake Victoria Fisheries Research Project, funded by the European Development Fund of the European Union (Project No. 7 RPR 372), whose support is gratefully acknowledged. This paper has benefited from comments and suggestions made by E. Mlahagwa, M. Kabati, D. Ngusa, D. Msunga, J. Ouko, J. Onyango, A. Asila, E. Allison and M.-T. Sarch.

9

Lake Victoria Fishers' Attitudes Towards Management and Co-Management

Douglas C. Wilson

Fisheries co-management programmes are appearing throughout Africa, particularly with regard to inland fisheries. Co-management has been argued to increase the effectiveness of management through increasing the legitimacy of fishing regulations, reducing and managing conflicts, improving information flows, and enhancing rule enforcement (Jentoft, 1989; Pinkerton, 1989). Current research focusing on the conditions under which these promises are delivered has found that both pre-existing social conditions, and how and by whom co-management is introduced, are critical, particularly in relation to the amount and type of conflicts over the resource (Normann *et al.*, 1998). The term 'co-management' refers to many different arrangements of varying structures and qualities that involve co-operation between government fisheries agencies and the fishing industry broadly understood (Sen and Nielsen, 1996). This definition is helpful if the analytic task is to catalogue and understand the various forms that such co-operation can take. It has been argued, however, that this broad definition diminishes the critical thrust of the co-management idea by including arrangements that do not involve real community-level decision-making power nor tap into the conceptual resources promised by the effective inclusion of the fishers' view of the resource (Holm *et al.*, 2000; Jentoft *et al.*, 1998). For this reason, the present chapter uses the term 'community-based co-management,' meaning programmes that involve substantial local decision-making power (a co-operative, advisory or informative type of co-management programme in Sen and Nielsen's (1996) terms) based on democratically organised community or industry-level decision-making institutions.

The research reported on here was carried out on the shores of Lake Victoria in Tanzania. Co-management was not in place at the time of the research, but was being contemplated. Possibilities for alternative management approaches were an important backdrop to the design and execution of the research, which both directly and indirectly addressed factors in the lake shore communities that would affect the success of co-management among other subjects. The research team talked to and surveyed fishers and other community members about four management-related topics. The first was their understanding of the reasons for changes in fish stocks. The second was their perceptions of those responsible for managing the fishery, i.e., the fisheries officers who establish and enforce management measures, and the scientists who identify what measures are needed. Third, we asked their thoughts on specific measures. Finally, we were interested in gauging their feelings about the idea of co-management.

Co-management efforts have now begun in Tanzania in the sense of engaging fishers in surveillance and enforcement efforts. These events reflect stated government policy that specifically encourages communities to 'conduct joint surveillance and enforcement in collaboration with other related agencies and communities' (Geheb 2000a: 179) and have taken practical steps with the introduction of 93 'local enforcement units,' in 1998. These groups were later renamed Beach Management Units and are incorporated into village governments as part of the village committees for surveillance and security (Geheb, 2000a).

Methods

This research was part of a larger study by the Programme on the Lakes of East Africa that was focused on gathering basic human ecological knowledge about Lakes Malawi and Victoria. At the point this programme was created, very little systematic research had been done on the characteristics, attitudes and activities of the people who were making use of these lakes. The research strategy was to combine quantitative and qualitative approaches by nesting the quantitative work into a broader qualitative strategy. Qualitative work was done to discover the important questions that needed to be asked, to ask these questions to different people in different ways, and to design questions for the random sample survey[48]. That survey then allowed us to both deepen our knowledge, and to evaluate the breadth and distribution of the qualitative information being gathered.

The data on fishers' attitudes towards management were generated by an in-depth, multi-method investigation of Lake Victoria's fishing industry. The data were gathered in visits to lakeside communities done in two phases: from January to July of 1993 on six fishing beaches; and from June to November of 1994 on an additional five beaches. We also did follow-up visits on the first six beaches during 1994. A survey was conducted with stratified random

---

[48] A full description of the research methods, as well as related publications and reports are available at http://www.isp.msu.edu/AfricanStudies/Plea/pleatop.htm

samples of boat management (n=145) and crew (n=241), lake shore households (n=178), and fish processors and traders (n=95). In addition, participant observation, focus groups and in-depth interviews were done with the same populations during the approximately two weeks that the research team stayed on each beach.

We defined a fishing beach as any place where fishing boats gathered on shore and where fish were traded. The beaches were selected randomly within two strata: the number of boats that landed fish on the beach and five ecological zones (Kudhongania and Cordone, 1974) that ensured both geographical and limnological spread. Fisheries Department census data were used to randomly select 15 beaches, one large beach where more than 30 boats landed fish, one medium sized 20-30 boat beach and one smaller beach were chosen in each ecological zone. Because of resource limitations, data gathering was completed on only nine of these beaches. When we arrived on a beach, we did a census of the fishing boats. If there were fewer than 25 fishing boats we selected all of them for our sample; if there were more we randomly selected 25. The owner, operator (skipper) if this was a different person, and randomly selected crew members were interviewed from each selected boat.

The Lake Victoria fishery

Lake Victoria, shared by Kenya, Uganda and Tanzania, is the second largest freshwater lake in the world. The lake is relatively shallow and one result of this is that it is an extremely rich source of nutrition. An exotic species, the piscivorus Nile perch (*Lates niloticus*), was introduced to Lake Victoria in the 1950s and experienced an explosive growth in population in the late 1970s. Because of this introduction, in combination with fishing pressure and eutrophication stemming mainly from runoff (Bundy and Pitcher, 1995), the lake was transformed from a multi-species lake to one in which there are three commercially important species: Nile perch, Nile tilapia, (*Oreochromis niloticus*), and the sardine-like *dagaa* (*Rastrineobola argentea*). The Nile perch is a large, white, meaty fish that finds a ready international market and an industrial processing and export industry grew up in Kenya and Uganda during the 1980s and in Tanzania in the early 1990s (Reynolds *et al.*, 1992).

In Tanzania, there are four major types of fishing units on the lake: Nile perch gill nets (49% of our sample) and long lines (9%), boats targeting *dagaa* (23%) and beach seines (9%). Nile perch are fished with gill nets and secondarily with multi-hook long lines. Gill nets are like huge tennis nets suspended in the water and catch the fish because they get their gills stuck in the net. *Dagaa* are fished at night when the moon is dark using Coleman-type pressure lamps to attract them. They are caught with several types of gear. Short purse seines and mosquito nets are the most common (62% of *dagaa* units in our sample) but scoop nets (35%) are also used. Beach seines are long nets (usually 100-300 metres but they can be much longer) that have a stocking-shaped central area with a mesh of about a half-inch. The boat leaves one end of the seine on the beach, places the net in a wide circle, finally returning to the beach where both ends are pulled in to shore. The gill net fishery is the one that is the most involved in the international trade. Gill nets are the dominant gear on the beaches that are more centrally involved in this trade but not on those that are more isolated. In our sample, 71% of the boats on central beaches are gill net boats while they make up only 32% of the boats on isolated beaches (central and isolated are quantitatively defined below).

Causes of changes in catch size

In the survey, fishers who had fished a species for five years were asked to compare their current and past catches. These comparisons (Table 9) and informal conversations found a general perception of declining catches.

Change	Nile perch	Tilapia	*Dagaa*	Total
Many more	11	8	13	11
More	11	8	15	12
The same	6	0	18	8
Fewer	50	69	51	51
Many fewer	22	15	3	18
Total 5 year fishers	156	14	39	206

Table 9: Changes in perceived catch rates – percentages Table reports percentage of those that had fished the same species for five years comparing current catches with those of five years ago. Two Nile perch and one tilapia fisher are missing due to error. These lost data errors, which are reported in all tables, resulted from interviewer or data entry error or, more commonly, from a number of interview pages and sections that were physically lost during the research.

In group interviews, people disagreed about both the causes of changes in catch rates and whether or not anything could be done about them. A small number gave various religious explanations for the decline, others pointed at natural causes and many saw humans as the prime agent. Drought is a very common explanation for declining catches, especially by fishers at the southern end of Mwanza Gulf. In this long, narrow, southern extension of the lake, changes in the lake level caused by an extended period of low rainfall were believed to have contributed to Nile perch catch declines. They argued that because the shallow water was muddier and thus less attractive to the perch, it had given the haplochromis, catfish and lungfish a chance to rebound. Participants in several focus groups also blamed the then growing water hyacinth infestation for depressing catches. The fishers reported that the water hyacinth covers areas that had been good breeding grounds.

Fishers and traders who worry about having enough fish in the future and those who do not worry were split almost evenly: 47% of respondents do not worry while 53% respondents do worry. For Nile perch fishers, this variable is significantly related to perceived past changes as can be seen in Table 10.

The who are worried were asked why they expected lower future catches (Table 11). Most often they cited changes in the weather as the primary reason. Weather is the clearest English translation of the Swahili expression '*hali ya hewa*' or 'condition of the air' but it does not capture the sense very well.

		Greater or unchanged catch compared with five years ago	**Lower catch than five years ago**
Worry about having enough fish in future?	No	58	40
	Yes	42	60
	N	43	110
	P	0.06	

Table 10: Nile perch fishers who worry about the future by perceived changes in catch – percentages Table reports percentages of five year Nile perch fishers only, five respondents are missing due to error.

The changes to which the fishers are pointing with this expression are general changes in the physical conditions of the lake and the environment which affect both numbers of fish and the ease of fishing. Several times they specified drought as the main cause, and some also talked about changes in wind patterns which make fishing more difficult. A group of destructive fishing practices were mentioned by different fishers: fishing in breeding grounds, fishing for juveniles, the use of pesticides for fishing, and beating the water to drive fish into nets.

Reason	% fishers mentioning this reason
Changes in weather	41
Over-fishing	21
Too many fishers	11
Fishing for juveniles	9
Predation by Nile perch	7
Too much gear	7
Fishing in breeding grounds	5

Table 11: Reasons offered for expected lack of fish by those who are worried about having enough fish in the future – percentages Table reports responses to an open-ended (i.e. there were no suggested responses) follow-up request for reasons from those respondents who said that they were worried about having enough fish in the future. These data are for fishers only, not fishers and traders. Each fisher was given the opportunity to mention up to three reasons; 193 fishers mentioned 20 reasons, 16 were mentioned more than once. Only those mentioned by five per cent or more are reported here. The filter question about degree of worry has four missing due to error and one don't know (DK) respondent.

Many of those who are worried about future catches recognize variables that can be manipulated by management as being important to the fish stock and their catch. Only two – changes in weather and Nile perch predation – of the seven most cited reasons cannot, in principle, be addressed by a management regime.

Those who do not worry about the number of fish in the future cite most often the rapid reproduction of the Nile perch as the reason (Table 12). Several said that people have been fishing for a long time without depleting the stock. 5% said that the drop in catch being experienced was seasonal. Others expressed the opinion that there are actually few fishers compared to the size of the lake and that they are using primitive gears not capable of finishing the fish. Two fishers said they were not worried because if the fish became depleted 'they' would just put in another species. Many fishers are fairly sophisticated about the biology of the lake: Nile perch are indeed highly fecund. Some are less sophisticated. We were once told that since the Nile perch came so suddenly they will probably leave just as suddenly.

Reason	% of fishers mentioning this reason
High fecundity of Nile Perch	62
There are many fish or the lake is very big	19
There is little fishing pressure or there are few fishers	8
Changes in catches are seasonal	5

Table 12: Reasons given for not worrying by those who do not worry about having enough fish in the future – percentages Table reports responses to an open ended follow-up request for reasons from those respondents who said that they were not worried about having enough fish in the future. These data are for fishers only. The number of reasons that fishers could offer was not restrained. 199 fishers mentioned 12 reasons, 11 were mentioned more than once. Table reports reasons mentioned by five per-cent or more of the fishers. The filter question about degree of worry has four missing and one don't know (DK) respondent.

Perceptions of the government role in management

On a general level, focus group discussions indicated that fishers consider fisheries management a legitimate activity. They see it as necessary and they believe that the rules are designed to ensure that there will be enough fish in the future. They also believe that the rules are established, though not necessarily enforced, for the benefit of all the users. On the whole 'science' is perceived as a source of valid truth. Fisheries scientists know what they are talking about because they have been educated and 'it is their job.' Scientists are viewed as honest and given respect.

Tanzanian fisheries management involves two government agencies, the Tanzanian Fisheries Research Institute (TAFIRI) and the Tanzanian Fisheries Division (TFD). TAFIRI is responsible for research and providing biological and socio-economic information about the fisheries while the TFD is responsible for the direct fisheries management activities. Several of the beaches we visited had agents of the TFD, referred to locally as '*Bwana Samaki*' (Mr Fish). One open-ended question asked 'when you want help or advice concerning fishing or the fish business who do you ask?' 33% of the fishers mentioned the *Bwana Samaki*. This was more than any other person, including friends and relatives (28%), no one (18%) and their employer (15%). This question was part of the section of the interview dealing with fisheries management issues, however, so the overall discursive context included the idea of fisheries officers.

The presence of fisheries officers was found to vary widely by beach. Table 13 describes these differences.

Beach Location	**Musoma Region (E. shore)**		**Eastern Mwanza Region (SE Shore)**		**Western Mwanza Region (SW Shore) (SW Shore)**		**Kagera Region (W. Shore)**		
Beach size	Medium	Medium	Large	Small	Small	Small	Medium	Medium	Large
Every day	0	100	87	53	100	0	9	6	0
Once or twice a week	2	0	2	40	0	17	88	69	2
Once or twice a month	54	0	8	7	0	83	3	24	94
Once or twice a year	30	0	0	0	0	0	0	0	2
Never	14	0	3	0	0	0	0	0	2
N	50	60	63	15	24	12	64	36	56

Table 13: Frequency of seeing local fisheries officer on the beach – percentages Table reports percentages of fisher respondents to the question 'How often do you see fisheries officers carrying out their duties on you beach' N=380, 17 responses are missing due to error. Location information in the table is for the reader's orientation. These regions do not correspond to the ecological zone used for beach selection.

Those beaches that reported seeing officers frequently are those where officers are resident, as officers have neither funds nor means for travel by road or lake. These beaches tend to be larger and more accessible beaches, but this is not a fast pattern. Frequency of officers' presence does not affect the familiarity of fishers with the fisheries regulations. For 14 different possible regulations each fisher was asked if a regulation of this type existed on his beach. The local fisheries officers were asked the same question in order to establish a baseline 'correct' answer. The answers for a beach where 98% reported never seeing officers or seeing them only once or twice a month or year and those for a beach where 100% reported seeing the officer every day, revealed no significant differences in the number of fishers who disagreed with the fisheries officer or who answered 'I don't know'. This suggests that there is little active enforcement or education about the fisheries rules.

The fishers often expressed worry about corruption in respect to fisheries officers, particularly in their appointment. There was a feeling that the local officers' jobs were patronage jobs that were sometimes given to unqualified people. There were also frequent reports of corruption, such as bribes, in the enforcement process.

Attitudes toward specific rules[49]

The perception of the rules gleaned from the qualitative interviews is that the government, at least *prima facie*, knows best what they should be. This legitimacy is diminished when the government is seen as inconsistent in enforcement, corrupt or incompetent. The fishers' major concerns are with making some of the rules that are already in force effective and they want to see consistency in enforcement. It is those activities which they consider, with variable justification, to be bad fishing practices (poison, small mesh sizes, beating the water to drive fish into nets, etc), which they want to see enforced, much more than any restrictions on access to the lake. One question that was raised a number of times in the groups was why illegal gears were allowed to be sold, ignoring the fact that there are other lakes where such nets are legal.

In interviews, biologists at TAFIRI indicated that they were most concerned with four illegal fishing practices: beach seines which catch very small fish and disrupt spawning; trawling vessels (beach seines and trawlers were banned while our research was ongoing), the use of poison for fishing; and using traps and weirs on river mouths during the spawning runs of some of the minor species (see also Bwathondi, 1992). There are a number of closed seasons and areas on the books that generally apply to bays and river mouths. Fishing boats are licensed, which provides a possible mechanism for controlling effort. A minimum mesh size of 5 inches for Nile perch and 0.4 inches for *dagaa* is also in effect. Other management measures can also be found that have been imposed by more local fisheries department officers at the regional and district levels. We used the statements of the fisheries officer responsible for the beach as the final determinant of what was considered legal on a particular beach as this would be the most locally effective official judgment.

The openness of rule breaking varies from rule to rule. Catching juvenile Nile perch is illegal on some beaches but goes on without embarrassment. So does the use of mosquito nets for *dagaa*, when the law requires a minimum mesh size of 0.4 inches. On the other hand, fishing using poison is done at night and remains hidden from the other fishers. Poisoners are strongly disliked and poison fishing is regarded as a serious crime. The apprehension of poisoners is sometimes reported on the radio. The reason the fishers give is that it destroys the eggs as well as the juvenile fish. Drumming the water to frighten fish into nets is also frowned upon by many fishers and is usually done at night, although it is a noisy activity and the darkness does not hide the fact that it is going on even if it makes it more difficult to positively identify who is doing it. Fishing

---

[49] Some of the material presented here was also very briefly mentioned in Wilson *et al.* 1999 in the context of a description of policy discourses around the management of the lake. This section both expands on that material and adds new information.

for juveniles and in breeding areas were not considered good practices by most fishers, but they did have their defenders in people who did not think it made much difference.

One important reason why many rules are broken with no attempt to hide the activities is that awareness of rules that are in place is not high. Only 7% of all fishers said that they were aware that there is any regulation concerning mesh sizes. 11% of the fishers in Kagera Region were aware of the newly promulgated rule banning beach seines. Only 26% of the fishers on beaches with a closed seasons rule (i.e. those on or near a bay or river mouth), responded affirmatively when asked if there was one. Awareness increases, of course, when there is enforcement. The 26% that were aware of a closed season rule rises to 35% when restricted to boat owners and renters and to 70% of the owners and renters when restricted to Mwanza Gulf where enforcement was more active. Boat owners and renters were more aware of regulations partly because their responsibilities required them to be, but also because the crew tended to be more transient both between fisheries and between fishing and other activities. Most decisions that involved rule compliance (e.g. what gear to use, when to go fishing) were actually made on the beach by the owners and renters.

Compliance with closed seasons was variable. What we observed on the Mwanza Gulf, which is considered a bay under the closed season rule, was that the number of fishers in these areas would decrease in the off-season, but that there would still be many boats fishing. One hypothesis is that those fishers who are mobile enough or close enough to get to the main lake during the closed season do so, while the others ignore the ban. This would indicate that officers were more indulgent of infractions by the less mobile group, or that paying bribes in order to not move was considered more worthwhile by some boats than others. Probably both factors were operating. On the beach on the southern end of Mwanza Gulf, which was far from the main lake, this regulation was described as 'just words'. On another isolated beach where, weather aside, the local bay was no more or less accessible than the open lake, a few fishers considered the regulation an issue in deciding where to fish.

One belief that was strongly held by many people in the group interviews was that any Tanzanian had a right to fish in Lake Victoria regardless of their original home. In one group interview, a participant said that 'in fishing there is no segregation' using a Swahili word, '*ubaguzi,*' which has strong historical overtones from colonial abuses and the fight for independence. For three different kinds of limits on access, an average of 72% of respondents rated them negatively or very negatively, with an average of 45% rating them very negatively (Table 14).

Management Measure	Percentage evaluating measures Very Positively	Positively	Very Negatively	Negatively	N
Limiting number of people allowed to fish.	12	16	24	48	201
Limiting number of nets allowed to fish.	16	17	21	45	201
Limiting number of boats allowed to fish.	10	12	35	43	201
Banning beach seines.	20	29	34	17	200
Banning mosquito nets.	30	37	27	6	201
Banning fishing in bays /river mouths from Jan. - June.	40	41	13	6	199
Banning gill nets with mesh sizes < 5"	35	39	21	5	199

Table 14: Evaluation of fisheries management measures - percentages These questions were asked only of fishers and were added to the survey in the second year, they were not asked at all on three beaches and only of boat owners, renters and operators on three others during follow up visits. For N = 201, thirty responses are missing due to error and four responses were neutral or DK, for N=200 is one more neutral response and N=199 is two more missing due to error.

This commitment to open access also stemmed from the fishers' own desire to be able to move freely to follow fish, prices, and security. Table 14 suggests that fishers were more inclined to regard positively those management measures which were already in place, a distinction which was indicated when the questions were asked: 81% rated the closed seasons in bays positively and 74% did so for the current ban on gill nets of less than five inches. Controls on mesh sizes were, however, generally considered a good management measure. 67% favored the banning of mosquito seines while opinion on the banning of beach seines was split with 49% for and 51% against. Dichotomous questions about whether or not there should be a minimum mesh size on gill nets, beach seines, and *dagaa* nets all received less than 15% negative answers – only one respondent in the survey was opposed to any mesh size restrictions on gill nets. Respondents were also asked to rank five fishing practices from least to most destructive, with 'destructive' being translated as 'bad for future catches' (Table 15). Beach seines were ranked as the least destructive by a wide margin. Fishing in spawning areas and using small mesh sizes formed a second level of medium ranking. Catching juvenile fish and beating the water to drive fish into nets were ranked as the worst offences. These latter practices were usually officially banned, but practised anyway.

Beach seines	0.93
Fishing in spawning areas	2.04
Using small mesh sizes	2.04
Catching juvenile fish	2.46
Beating the water to drive fish into nets	2.53

Table 15: Mean rank ordering of fishing practices Fishers were asked to rank the five practices from the worst = 1 to least bad = 5. Reported values were inverted and subtracted from five for clarity of presentation. N=219 These questions were asked only of fishers and were added to the survey in the second year, they were not asked at all on three beaches and only of boat owners, renters and operators on three others during follow-up visits. Data from fourteen respondents are missing due to error, two responses are DK.

Beach seines were the most controversial issue. We found many differences of opinion around the lake about whether a beach seine with an average of 2.5 inch mesh size in its wings was really too small and could be blamed for depleting the stocks. Seine owners said that banning seines might improve fishing, but these improvements benefited only the '*wavuvi wa kisasa*' (up-to-date fishers), i.e., those who used engines and gill nets farther offshore. Beach seines played a critical role in the fishing communities. As we argue in Wilson *et al.* (1999), pullers of the seines were young men and boys who worked for '*mboga*,' i.e. fish for family consumption. Local families who did not have fishing gear had the option of sending their children down to the beach for a few hours or a night to pull the seine to bring home high quality protein. The attitude of the fishers in the Mwanza Gulf village to a large beach seine operation was instructive. On the one hand, they believed that this operation was over-fishing Nile perch, and felt that this was unfair to the rest of the village. On the other hand, they were reluctant to criticise a local family that had made good, and which played an important role in transporting the village's fish to market.

We also found management professionals who were not convinced that banning beach seines would make any significant biological difference. While they agreed that they are destructive where they are used, finding areas of the lake shore that are free enough of rocks and vegetation to use them is difficult. The beach seine ban was also partly driven by the international politics around Lake Victoria because it was the most easily enforced measure among those approved of by donors and NGOs (Wilson, 1999). In recent follow-up work, Medard (1996) found strong resistance to the ban. She looked at three beaches on Speke Gulf. 17 beach seines were present and being used at night, while enforcement was hampered by bribery. Those groups who had not been benefiting from the export market, particularly the small-scale fish traders, were the most supportive of the continued use of beach seines and other illegal, small-meshed nets.

Factor analysis is done to help identify underlying, independent attitudes that are revealed by answers to more than one question. Table 16 reports a factor analysis done with 10 variables measuring attitudes towards specific management measures.

Variables	**Factors - Varimax Rotation**			
Five point scale evaluating:	*1*	*2*	*3*	*4*
Limiting the number of fishers.	.79	.12	.05	.28
Limiting the number of nets.	.80	0	.01	.03
Limiting the number of boats.	.79	-.03	-.02	-.04
Banning beach seines.	.15	-.10	-.03	.88
Banning mosquito nets.	-.09	.80	0	.24
Closed seasons.	.04	-.08	-.86	.22
Banning gill nets of less than 5 inches	.08	.24	.71	.41
Yes or No Response to:				
Should there be a minimum mesh size for *dagaa* nets?	.06	.81	.10	-.12
Should there be a minimum mesh size for beach seines?	.10	.59	.22	-.21
Variance explained	22%	19%	11%	10%

Table 16: Factor loadings for management attitudes N= 173. Invalid responses to any question removes respondent from the factor analysis. Thirty responses to at least one question are missing due to error. Thirty-two respondents responded DK to at least one question, 26 of these were Nile perch fishers with no opinion about *dagaa* nets.

Factors are linear combinations of variables that do not correlate at all with each other and the factor loadings are the correlations between each variable and factor. Each respondent can also be assigned a score on each factor, which allows the calculation of correlations between the factors and variables not used in the factor analysis. All the attitude variables have been coded so that a higher score indicates a more positive attitude towards active managerial control of the activity in question.

The factor analysis found four factors that explained at least 10% of the variance of the original variables. The first factor clearly reflected an attitude toward restrictions on access. As was reported in Table 14, most fishers did not wish to see any restrictions on access. Its emergence as a factor here indicates that this is a single attitude directed at all three forms of access and it is an attitude that is independent of attitudes towards other specific management measures. The factor loadings are really too close. It was a methodological error to place three such similar questions next to each other in the interview schedule as the response to the first may bias responses to the other two.

The second factor revealed an underlying attitude related particularly to the control of the mesh sizes of beach seines and *dagaa* nets. Positive attitudes towards this kind of management were found on more isolated beaches (Table 17), meaning those beaches on which less than 30% of the fish sold in the week before the interviews was destined for the international market. It also correlates at -.34**[50] with the size of the fishing operations on which R works (Table 17), measured by total value of boats and gear owned by the owner of the selected boat.

Mean Factor Scores		**Mean Factor Scores** 1	2	3	4	N
Centrality of beach	Central	0	-.47	0	.37	53
	Isolated	0	.21	0	-.17	120
	Significance (P)	NS	0	NS	0	173
Fishery	Beach seine	-0.09	-0.05	-0.05	-0.43	19
	Nile perch gill net	-0.05	-0.17	0.17	0.26	72
	Dagaa	0.04	0.08	-0.12	-0.21	37
	Nile perch long	.12	-0.17	-0.12	.02	27
	Significance (P)	NS	NS	NS	.04	155
Correlation with size of operation		-.10	-.34**	-.04	.24**	170

Table 17: Distribution of factor scores Lower N for fishery results from fishers on eight boats with data missing by error and eight boats excluded because they fished for minor species. Lower N for size of operation results from data missing by error.

The third factor is the most ambiguous. It tells us that attitudes towards closed seasons tend to correlate negatively with attitudes towards whether or not there should be a minimum size of 5" on gill net mesh and that people holding these two attitudes do not show a pattern on the other variables. Factor three does not correlate with any other variables, save slight correlations with wanting to see control over mesh size of other types of nets. This result cannot be interpreted as a preference for gear restrictions over access control, as there is no correlation between attitude toward seasons and other indicators of attitudes towards access. Because the attitudes towards mesh restrictions load to some extent on factors two and four as well, the best interpretation of factor three is

[50] Throughout the text, statistical significance is indicated by: * = significant at .10; ** = significant at .05; and *** = significant at .01.

that some fishers share a positive regard for mesh restrictions with others, but have a particularly strong negative opinion toward closed seasons.

The fourth factor indicates that the attitude towards beach seines is also distinct from other attitudes expressed. The only attitude that loads at greater than a .5 is evaluation of the banning of beach seines. Scores on this factor are significantly higher on central beaches than on isolated ones and the factor correlates at .24** with the size of the fishing operation, both as defined above (Table 17). Table 18 provides some interesting detail as it reports the raw numbers underlying the relationships uncovered in Table 17.

Fishers, of course, resisted bans of their own gear but were not opposed to controls on mesh size. In fact, the data suggests, albeit without significance, that they were slightly more likely to support mesh size controls on their own gear than on other gear. Nile perch gill netters on central beaches supported a ban on beach seines while other groups are opposed.

Fishers on isolated beaches were more supportive of restrictions on beach seine and *dagaa* net mesh sizes, hence they supported mesh size control on gears that were used on the beaches from which they fished. This contrasts with the suggestion of slightly lower support for these measures from Nile perch fishers. Overall, fishers' attitudes toward management reflect both their own operations and the types of operations being used in the community around them.

Attitudes towards co-management

We also specifically addressed the idea of co-management and the fishers' reaction to it. Our standard introduction to the subject ran as follows: 'Some people think the fishery would benefit from local fishers controlling who can or cannot fish, what gear can or cannot be used, and where and when fishers can or cannot fish. People suggest that this might be done through a co-operative or some other organization where the fishers can work together. We would like to know your opinion.' Our emphasis, then, was on increased community control of fisheries management decision-making.

		Supports banning beach seines			**Supports controlling beach seine mesh**			**Supports controlling *dagaa* net mesh**		
	Overall	Yes %	No %	N	Yes %	No %	N	Yes %	No %	N
		49	51	200	92	8	185	88	12	179
Fishery	Beach seine	25	75	20	95	5	21	85	15	20
	NP Gill net	60	40	91	86	14	76	84	16	74
	Dagaa	43	57	40	95	5	38	92	8	39
	Long Line	38	62	29	97	3	30	89	11	28
	p		0			NS			NS	
Beach type	Isolated	36	64	128	97	3	128	94	6	123
	Central	71	29	73	81	19	57	73	27	56
	p		0			0			0	

Table 18: Fishers supporting selected management measures – percentages Table reports responses to one survey question that rated the measure 'to ban beach seines' on a five point scale from a very good to a very bad rule and two survey questions which answers yes or no to 'in your opinion should there be a minimum mesh size for nets targeting *dagaa* or beach seines.' These questions were asked of fishers only. They were added to the survey in the second year and asked only of management during follow-up visits to four of the six first-year beaches. As a result, management is over-represented in this sample by 12% compared to the overall sample. N of 200 represents 43 missing responses and 5 DK, 185 is 43 and 20 respectively and 179 is 46 and 23 respectively. Lower Ns on fishery variable reflect exclusion of 11 boats with minor uses, 1 boat that had not yet fished and 8 boats for which operations data is lost for columns one and two and 10, 1, and 7 respectively for column three.

The responses we received reflected the fishers' own experiences with how fishing communities were organized and operating. Current local community organisation was made up of four main groups. The importance of the groups varied from beach to beach. The most important group when we were present were the local officers of the Chama cha Mapinduzi (CCM), the ruling party since independence. Since the 1960s, the local party organization of a Chairman, a Secretary, and a group of '*Mabalozi*', or Ten Cell Leaders, have acted as the centre of village government and we found these Ten Cell Leaders even in the temporary fishing camps. The opening up of the national political process has forced them to compete with other political groups but in the villages we visited, the CCM continued to handle local functions. The Tanzanian Government's Provincial Administration is present, but functions on a slightly higher geographical scale, leaving it to the CCM to provide day-to-day leadership at the lowest levels. The third important leg of local organization is the *sungusungu*. These are citizens' groups that are present in every village and act as local police forces, but with little constraint from legal formalities. The local leader of the *sungusungu* is an important local personage.

Beyond these three types of official leadership, most of the beaches have some sort of organised leadership among the fishers themselves lead by a 'beach leader.' The interplay between the official leadership and the way fishers are organized differed at every beach we visited. On the two beaches we visited that were 'fish camps', meaning relatively temporary communities made up of migratory fishers, the local organisation could be described as 'government by boat owner'. All of the people in positions of authority were boat owners; and, in both cases, authority was anchored through negotiations in which fishing operations were the key identity. In an interesting illustration of how this conflation of boat owner and village government was perceived, the 'village' government of the largest fish camp was for all intents and purposes the beach leader. Crew members in that camp reported that selling fish on the lake was a violation of village law, the punishment for which was loss of employment.

Security from gear theft was the focus of the most active beach-level organising, though this was still only happening on a few beaches. Theft of nets was considered the most important problem by the fishers. Fully 70% of fishers who had been fishing for five years had had a boat or some sort of fishing gear stolen from them during that time, and 22% of these five-year fishers had been victims more than once (Table 19).

Number of Times a Victim in Last Five Years	Percentage
0	30
1	48
2	12
3	7
4	3
N	121

Table 19: Theft victimization among five-year fishers who own gear Table reports responses from only those fishers who both have been fishing for five or more years and who own or had owned boats or gear.

After following the fish and finding better prices, security was the third most common reason cited by migratory fishers for deciding where to move. During a group interview with village elders, the focus of which was the area's history, the subject kept returning to the theft of nets. Nearly every male present indicated that being a victim of gear theft had precipitated his leaving fishing. On two beaches we visited, there had been serious attempts to create community-level institutions to address the security problem. On one, an active fisheries officer has been an important agent in bringing about regular patrols. Their plan was to organise seven to ten boats without engines into groups with a boat that did have an engine. All the fishers then shared the costs of having the engine boat patrol the gill nets at night. On the other beach, the patrols were done by one boat with an engine that charged one-seventh of the total catch for the service.

While reactions to the idea of fishers participating in the management of the lake fisheries were mainly positive, many reservations were also expressed. Fishers' understandings of how local control of fishing might operate reflected their perceptions of what were the current important problems and operating organisational practices. The models of co-management they expressed reflected the operations of the fisheries officers and/or the *sungusungu*. Fishers immediately jumped on the co-management idea as a way to reduce the threat of theft. The model they had in mind was adopting for their beach some co-operative patrols that were already operating on a small number of the beaches.

A number of doubts about organising for co-management and mutual protection were expressed. The size of the lake and the inability to know what other areas are doing generated questions about co-ordination. They wanted to know how co-management attempts in a small area could do any good if others are not doing it. Leadership was also seen as a problem. It was hard to find non-corrupt leaders. In the past, co-operative attempts failed because of a lack of leadership.

Particularly on the largest permanent beach we visited, one close to the city of Mwanza, fishers felt that they would need organisational and material help from the government if they were to be successful. Seminars and training were mentioned often, as were the supply of newer gears as an incentive to co-operation. This beach had a very active fisheries officer and this seems to have translated into more dependency on the government for organisation. On the smaller and more isolated beaches, this kind of government aid was mentioned rarely or not at all.

Another fear, that was expressed almost exclusively on this large permanent beach, was that co-management would be good for boat and net owners but not for crew. They feared that it would become a sort of boat owners' union that would reduce the crew members' bargaining power. An organisation controlled by the owners might make it difficult for the crew to offer their services to different boats.

Fishers said that everyone would have to make some sort of investment in co-management for it to work, yet they predicted that a large number would be reluctant to do so. Many said that it would be difficult because they are used to watching out for themselves, especially 'these days'. Thieves and those who lacked faith in the organisation would break it down. The migration of boats from beach to beach was also suggested as a source of organisational difficulties.

Conclusion: implications for attitudes for community based co-management

This combination of qualitative and quantitative methods has allowed the assessment of both what the attitudes are that have implications for co-management and how these attitudes are distributed in lake-side communities. The attitudes analysed here point to some of the ways that co-management would receive support within the fishing communities and these ways do not reflect a strongly community-based approach. This is more of a function of the models available, however, than a rejection of the idea itself, which is simply unfamiliar. Many fishers are interested in management and ready to co-operate with a government co-management programme. They are supportive of the idea of fisheries management and see many of the variables that can be affected by management as being important for the fisheries' future. In spite of this, most current management measures and institutions have little real substance on the beaches. Most fishers are not even aware of the management measures because they have little practical importance in their lives. An important job needs to be done to increase the limited awareness of management measures and issues through substantial community involvement in both education and enforcement.

As the complexity of the lake ecosystem varies greatly from place to place, the potential contribution of fishers to the management knowledge base also varies. The data on perceptions of causes of changes in catch size indicate that many fishers are sophisticated about the biological basis of management while others are not. In those areas away from the main lake where the Nile perch has not gained complete dominance, the fishers have particular local knowledge that is potentially useful to management. The fishery at the southern end of Mwanza Gulf, for example, is based on an idiosyncratic set of species that central authorities have no good means of monitoring. For the Nile perch, a more important contribution would be to gather the standard biological data needed for stock assessments, which are currently very haphazard. This fishery is a large-scale resource that will continue to require aggregated data to monitor.

Community-based organisations related to issues of both the fishery and co-operation with government regulations are certainly not alien to the existing model of beach operations. The beach leader system, the CCM and the *sungusungu* are all examples of strong community level organizations. It is easy to imagine community based co-management taking the form of a 'fisheries *sungusungu*' because this is a model that is currently very familiar in the fishing communities. This model of co-management as essentially a regulation-enforcement mechanism is attractive to both the fishers and the government. The government sees enforcement as an almost overwhelming problem given their current limited resources, and sees co-management as a potential source of volunteer labour. The fishers assign critical importance to security from theft and this would be an important, initially probably the most important, motivation for them to support a co-management programme. The organisation of the Beach Management Units described by Geheb (2000a) indicates that this is indeed the model that is emerging as Tanzania begins to move toward of a form of co-management. Local co-management groups themselves have been known to place great emphasis on their enforcement role. In Zambia and Malawi, where fisheries co-management programmes have been on the ground since the early 1990s, the Village Co-management Committees drive a constant debate over the level of policing authority that they should have available to them (pers. obs.)

The analysis of the distribution of attitudes toward management (Tables 16, 17 and 18) shows real disagreements about, and different perceptions of, management problems. As in all fisheries, the Lake Victoria fisheries contain both active and potential conflicts that would affect any co-management programme. As is so often the case, fishers do not want gear bans or other access restrictions on their activities, but they are quite willing to consider technical measures such as regulation of mesh sizes. Beyond this, observable divisions exist over management options, with the now banned, but still used, beach seines being the most important source of controversy. The attitude questions and factor analysis reveal systematic differences in perceptions of management issues, mainly between the gill net fishers, who tend to be more oriented toward the international market, and those using other gears, who tend to be oriented toward more local markets. These differences reflect conflicting interests stemming from relationships at higher scales, from the level of the lake (Wilson *et al.,* 1999) to the international level (Wilson, 1999).

A related phenomenon is that some fishers are more able and interested in engaging in management discussions than others. Those with more interest and awareness of management issues reflect certain groups more than others. They tend to be boat owners more then crew, fishers fishing for the export industry more than others, and migratory fishers more than sedentary ones

(Wilson *et al.*, 1999). This implies that any community-based co-management effort will face a danger of some interests being better represented than others.

Community-based co-management could help address these differences and dangers if the government took on the critical role of ensuring that empowered, community-level co-management groups are authentically representative of the competing interests in villages. Given that several local interests reflect the broader interests of the export industry, however, it may prove very difficult for the government to avoid giving greater weight to those interests (Wilson *et al.*, 1999). This would require both a willingness to devolve decision-making on the part of the government and an effort to convince the fishers, who are much more familiar with top-down models of decision-making than they are community-based ones. The government and any other party interested in developing an effective community-based co-management programme are going to have to manage a delicate balancing act between working with the local people most able to make co-management work, and still effectively including the broader lake-side community.

Acknowledgments

The author gratefully acknowledges the support of: the U.S. Department of Education's Fulbright-Hays Doctoral Dissertation Research Abroad Program; the John T. and Catherine D. MacArthur Foundation; the Michigan State University African Studies Center; National Science Foundation Dissertation Improvement Grant INT 9320235; the Tanzania Fisheries Research Institute; the Social Science Research Council International Predissertation Fellowship Program; and the ICLARM/IFM Fisheries Co-management Research Project. A great debt of thanks is owed to Joyce Frederick, Modesta Medard, Ramadhani Mhekela, and Bellarmine Zenge for perseverance in data gathering and Charles and Olivia Mkumbo for the same in data entry.

10

Conflicts amongst Resource Users: The Case of Kabangaja Fishing and Farming Community on Lake Victoria (Tanzania)

Modesta Medard, Kim Geheb and Joash B. Okeyo-Owuor

Conflict is used to designate any relationship between opposing forces, whether marked by violence or not. The word encompasses not only the manifest aspects of the opposing forces but the underlying tension between them. Conflicts originate in the different perceptions of the parties involved over who should manage, use and benefit from a resource (Desloges, 1997: 34). Ultimately, the reason why conflicts over resource use should arise is as a result of competing claims over a resource. These become accentuated particularly if the resource is scarce and claimants to the resource are many.

Conflicts over the appropriation and management of common property resources can pose significant problems for the creation of sustainable management strategies. Often, there are long-standing conflicts between government and private sector, as well as among and within communities, over fisheries resources, their use and management. With the emergence of trade liberalisation and the globalisation of economies, fisheries resources are coming under increasing pressure from a growing number of actors. Often, these are more powerful than communities both in terms of the investments they can make and the harvesting technology they are able to employ. With increasing demands on a decreasing resource base, the number of conflicts within communities themselves are also on the increase, creating mistrust between communities, and hampering the introduction of participatory methods of resource management.

Conversely, if community managerial action turns upon a commonly perceived dilemma – such as resource claims made by outside agencies – then management systems may well arise (Wilson, 1982). It is common throughout the small-scale fisheries of the world that management systems developed at the community level seem to be more preoccupied with minimising the potential for conflict between groups of resource users, than they are with conserving the resource *per se* (cf. Alexander, 1977; Forman, 1967; Levieil and Qrlove, 1990).

Alternatively, efforts to resolve conflicts may originate from beyond those in conflict. Such external interventions can, on the one hand, mitigate some of these conflicts but, on the other hand, can also exacerbate them and even create new ones.

This paper explores the conflict between two ethnic groups in a fishery: migrant Ha fishermen who have settled on the shores of Lake Victoria; and resident Sukuma people. In Tanzania, changes have been made to the formal managerial structure of Lake Victoria's fisheries which have introduced so-called Beach Management Units (BMUs) to most fishing communities around Tanzania's sector of the lake. These have been seen as a way of delegating some responsibilities for the management of the resource to communities. The BMU strategy of fisheries regulation expects fishing communities to enforce government fisheries regulations. As such, there is no community involvement in the design of the regulations which their BMUs are charged with enforcing. It is this external intervention that has opened up old wounds in the competing claims over the resource base at the study site examined in this chapter, Kabangaja.

The conflicts that permeate Kabangaja's communities are multi-dimensional (as many resource-use conflicts are), and are based on land (space), fisheries management, economic and socio-cultural factors. These problems have impeded efforts to implement the BMU managerial strategy and, conversely, the introduction of the BMU at Kabangaja has exacerbated the community's internal conflicts.

Considering the importance of these issues and the lack of information on the linkages between fisheries policies and conflicts, this chapter concludes that it is important for the fishing industry in Tanzania to closely examine these conflicts. They are crucial in the understanding of the social processes that may one day generate the sustainable management of the fishery. Conflict should not be viewed as the dysfunctional relationship between resources user groups to be avoided at all costs, but as the constructive change and growth of society.

Research methods and sample selection

This paper draws on interviews with 33 fishers at Kabangaja beach, carried out in November 1999. Two focus groups with fishers and two with elders and village leaders from the farming community were also held during follow-up visits in August and December 2000. Respondents were selected randomly from a list of the beach's fishers. Respondents for the focus group discussions were obtained using stratified sampling procedures. Seven to ten people over 18 years old were invited to attend the discussions. In-depth interviews with influential actors within the community were also carried out. To strengthen the relationship between the two sampling techniques, some of the questions posed to fishers were also posed to the focus groups.

Community members' origins

Kabangaja is one of nine village cells at Igombe village in Mwanza Rural District (Figure 24). Igombe has a population of almost 5,000 while Kabangaja has a population of about 1,000. Of these, 224 are fishers based along the shoreline of the village cell. 215 of the fishers are thought to target *dagaa,* a small, endemic cyprinid fish species that comprised about 3% by weight of the fish landings from Tanzanian's Lake Victoria fishery in 1995. The remaining fishers comprise beach seiners, gill netters and long liners.

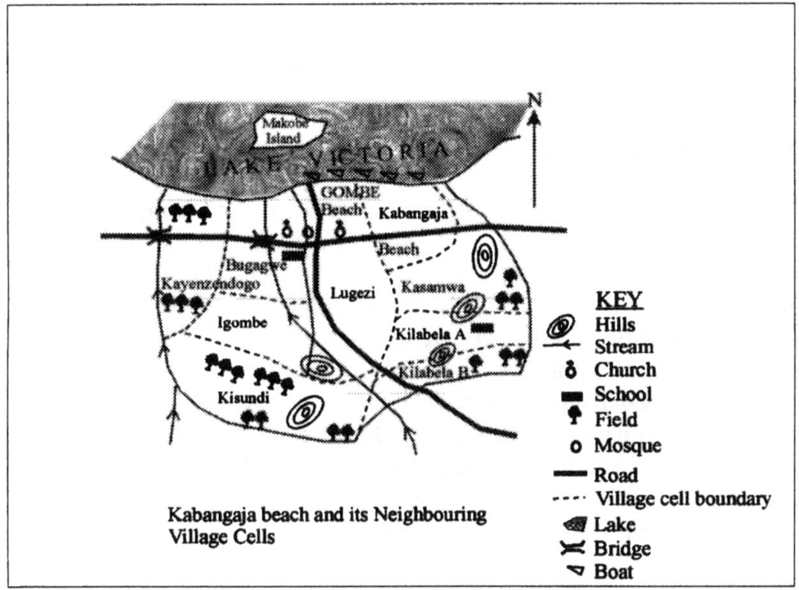

Kabangaja beach and its Neighbouring Village Cells

Figure 24: Kabangaja Cell and its neighbouring village cells

The name 'Kabangaja' is derived from the Sukuma word '*kenganza*' meaning a forest with lions in it, which used to occupy Kabangaja's present site. After the village was established, the situation changed. The forest disappeared, and soon there were only water reeds along the shoreline, which the fishers used to build their huts. In turn, the reeds were to disappear. The Sukuma are not, by this measure, 'natural' conservationists. A largely pastoral peoples, their practice of over-grazing grassland in an effort to control ticks (which inhabit various grassland shrubs, where they await passing livestock), has been widely commented on and is considered very destructive (cf. Birley, 1982; Charnley, 1997). Temple's (1965) examination of Sukuma land use practices revealed that his study subjects had a precise knowledge of the land and its soils. Temple concluded, however, that the land was being exhausted, as evinced by the use of marginal soils and the cultivation of crops able to survive on low nutrients. This was attributed to high levels of population growth, itself prompted by migrants seeking better soils because of soil exhaustion in their previous settlements.

All *dagaa* fishers in the study area were immigrants from Lake Tanganyika, who moved to Kabangaja with their fishing craft and crew by rail, settling along the shore in early 1989. Crew members were often related to one another, such that if one should return home, a relative would replace him. Most are Muslims. The fishers live in thatched huts made of mud, grass, boxes, blue UNHCR tarpaulin and polythene sheets. They are entirely dependent on fishing.

Major economic activities at Kabangaja

The main economic activities at Kabangaja are fishing and farming. A few community members keep goats, cows and sheep. During the dry season, these are left to roam free on the farms and other public land. During the rainy season, animals are grazed in specially designated areas, so that they do not damage the crop. The animals are watered at the lake, which the fishers dislike because their drying area is fouled with cow-dung.

The economy of households around Kabangaja was directly dependent on *dagaa* for sale or barter for other commodities. *Dagaa* is a freely available foodstuff for community members, and when the boats landed, women could be seen with bowls, pans and troughs going to the beach to ask for '*mboga*', a free fish supply sufficient for one to two day's worth of meals.

All those residing near the beach grow food and cash crops, but, in recent years, the majority of them could not rely entirely upon their fields to meet their food and financial needs due to the persistent drought. During this time, they relied on other seasonal income-generating activities such as selling peanuts, keeping food stalls and gaining incomes from seasonal labour migration. Labour included the building and construction of houses, brick-making, selling second-hand clothing, vegetable gardening, and cereal businesses.

Fishing provides higher returns to households than any other activity at Kabangaja. Fishers' craft and gear are never idle during the course of the year. If fishers are absent, their wives and relatives supervise fishing operations.

Sometimes, departing owners lend their equipment to close male friends, so that their families will still benefit from an income during their absence.

At Kabangaja, the *dagaa* fishery dominated until 1994, when fishers began targeting Haplochromines as well. The two species of fish are caught with the same gear, and in recent years have been harvested in almost equal quantities. Wholesalers from markets far from the lake will come to Kabangaja to buy large amounts of *dagaa* to be sent to regional and national towns, such as Mwanza, Arusha, Dar es Salaam and Mtwara, or even further away to the Democratic Republic of the Congo (DRC), Kenya, or Malawi.

At the beginning of 1994, an Italian investor, Polo Amadoli Italia, built a factory close to the fishers' settlement, with a view to collecting *dagaa* supplies to stock the company's animal feed factories. To the fishers' surprise, the prices offered by the firm were somewhat lower than those offered by buyers from distant markets. The relationship between the landing site and the firm were not good, and eventually, the investor diversified his supply base, and began seeking Nile perch by-products to use instead of *dagaa*.

Strengths and weaknesses of the BMU at Kabangaja

Kabangaja is served with several institutions located at the village cell government office and at Igombe village. It has a single primary school, a health centre, a primary judiciary court, a Ward government office and a Village government office. The village has three major committees which meet for day-to-day planning: the finance and planning committee, the peace and security committee and the community development committee. It also has an executive committee, which draws a chairperson from each of the nine village cells, along with other representatives.

The sub-village has two Fisheries Department (FD) staff working hand-in-hand with Local Government leaders at the Ward level. In July 1999, the FD held a meeting with village leaders and fishers to select Local Enforcement Units (LEUs), which were later to become Beach Management Units (BMUs). These are landing-based groups, drawn from members of the community, which are expected to enforce government fisheries regulations. The meeting was attended by almost all of Kabangaja's fishers and a few village members and officials from the village cell government. At the time, 10 resident community members and 10 *dagaa* fishers were selected to sit on the BMU. The FD stipulated their obligations as follows:

(a) To ensure the beach environment was clean.
(b) To avoid and prohibit the use of all illegal gears such as beach seines and undersized nets.
(c) To confiscate any illegal gear and report these to the FD.

(d) To ensure all new comers to Kabangaja were good fishers and that they reported to the BMU on arrival and thereafter to the FD for allocation of a camp site.
(e) To supervise fishing licensing at the beach level.

Beyond the physical energy of its members, the BMU did not, at first, have either the facilities or the resources on which to draw in order to fulfil its tasks. Because the fishers were used to receiving instructions directly from the FD, the creation of the BMU served to confuse matters, especially since the BMU was – as we shall see – crippled from the outset. This gave rise to different perceptions concerning the actual roles of the BMUs. Some thought that it would bring incomes to the people via salaries and allowances provided by the FD. Others thought that the BMU would be allowed to collect levies on catches as an incentive to carry out their duties. There were those who thought that the BMU would give them the power and independence necessary to resist interference from resident community members. Conversely, there were those who hoped that the BMU would be a unifying vehicle for Kabangaja's two communities.

By July/August 2000, the BMU had changed, and a completely new group had been selected without the knowledge of prominent *dagaa* fishers. Fishers did not understand why this had happened, or who was responsible for its dissolution. The proclamations made by the FD at the first BMU meeting - that fishers should be involved and consulted when it came to the management of the fishery - seemed not to apply to Kabangaja. The fishers wondered in what direction the FD was taking them, and whether or not the new BMU could be trusted. The fishers concluded that they had not been consulted because the government undervalued the *dagaa* fishery.

Fishers argued that theirs was a 'poor' fishery, providing *dagaa* for consumption to impoverished people, while the wealthy just used this fish to feed poultry. In view of the contribution they made to providing the poor with a cheap, readily available source of food, *dagaa* fishers felt that they deserved to be respected. Such respect, they felt, was not there. While the government worked hard to have the Nile perch export markets re-opened during an EU ban on these exports, they had done nothing to try and improve the markets for *dagaa*.

Such governmental attention, respondents argued, had implications for the strength and efficiency of the BMUs. At the neighbouring Nile perch beach of Igombe, fishers had greater responsibility and authority than did *dagaa* fishers at Kabangaja. Respondents claimed that while BMUs at Nile perch landings are supposed to record daily landings, no one had asked Kabangaja's fishers to do anything similar. At Nile perch landings, they complained, hygiene standards had improved so as to comply with foreign market demands, but no such improvements were forthcoming for *dagaa* fishers. Nile perch fishers and their landings had also benefited from road construction, and access to loans from Nile perch industrial processors. *Dagaa* landings, conversely, faced a number of problems such as poor storage facilities, filthy and sandy drying conditions and a lack of reliable markets. Gibbon (1997: 4) also points out that because *dagaa* is not exported to the northern hemisphere, and makes a relatively small contribution to the overall food security picture, neither the production nor marketing of the fish has ever been subjected to much government interest.

Kabangaja perspectives on management and its implementation

In November 1999, fishers at Kabangaja were asked who they thought should be involved in the monitoring of the lake. 51.5% replied that both the community and FD should be involved, 24% said that it should be the FD alone, while 24% said that it should be community members alone (N=33). Some fishers, however, were reluctant with the use of the word 'community', and qualified their responses by saying that if the term were used, then it should apply to *dagaa* fishers alone.

When the FD first attempted to establish the BMU at Kabangaja in 1999, it was intended that *dagaa* fishers and resident community members should be equally represented on the BMU. To their surprise, *dagaa* fishers found themselves a minority in the group. The new group was formed with six fishers, of which only two were *dagaa* fishers. The remaining six representatives on the BMU were all community members. *Dagaa* fishers argued that this was not satisfactory because some members from the new group were beach seiners and Nile perch fishers who did not represent the interests of the *dagaa* fishers dominating Kabangaja. They complained that the two *dagaa* fishers' representatives came from '...Kigoma interior, accepting everything and not capable of analysing issues because of lack of exposure' (Zeidi pers. comm.). Non-*dagaa* fishers, respondents claimed, were law-breakers at Kabangaja, and they therefore wanted to have more control than other fishers' groups. They argued that the fact that beach seining openly occurred at their landing was an indication of the BMU's weakness as a managerial institution.

Fishers were asked who they thought should be responsible for sanctioning offenders who had broken fishing rules and regulations. 27.2% said only the community should, 51.5% (N=17) said the Fisheries Department (FD) and 21.2% (N=33) said both the FD and the community should have this responsibility. The majority of the fishers interviewed did not think that offenders would be fairly punished if members from the resident community were involved in the process. This perspective was further reinforced when fishers were asked whether or not a local vigilante group, the *sungusungu*, should be involved in the management of Lake Victoria's fisheries. At Kabangaja, the *sungusungu* is wholly drawn from the ranks of the resident community. 75% of respondents did not think that the *sungusungu* should be involved in fisheries management activities, while the remainder did (N=33). *Dagaa* fishers interviewed claimed to have their own, secretly formed, *sungusungu* group, and if it was to this that the question referred, then they said that they would want it involved in fisheries management.

Fishers were asked what the community as a whole could do in order to ensure that community members changed from illegal to legal fishing practices. The following were mentioned:

(a) The old BMU should be restored. Respondents did not understand why a new BMU group had been formed, or what had been wrong with the original group. They did not understand under whose authority the new group was operating, and assumed that the BMU was non-existent at Kabangaja because their own needs were not being attended to.

(b) They argued that the FD should not change things without communicating with them. Because the FD had provided the new BMU with gumboots and sweaters, it was clear to them that the FD had accepted the new BMU, without, however, having consulted with the fishers.

(c) The BMU and the FD should, respondents felt, be fair when it came to monitoring fishing. When *dagaa* fishers wondered why it was that beach seining was going on, they were told it was not catching *dagaa* and they should not complain about it. This was demoralising and *dagaa* fishers were waiting to see the Fisheries Department's reaction.

About half of the respondents interviewed felt that only the FD should be allowed to determine which gears should be permitted on the lake. 31% felt that both the FD and the community should have the right to make this decision, while 19% felt that the community alone should have the right (N=32). Fishers, some argued, are the harvesters of the lake, and it does not, therefore, make sense to include non-harvesters in its management. During one focus group discussion, respondents raised the following points:

(a) Fishers should be more involved in the lake's management because management is geared towards solving day-to-day problems in their fishing operations. Having local fishers involved in management ensures that local problems can be solved which management for the lake as a whole probably could not deal with.

(b) Because the local residents do not get direct financial benefits from *dagaa* they have no feeling for the future management of the lake, unlike those who depend on fishing for food and income.

(c) The villagers are farmers. It makes no sense, therefore, to have farmers involved in fisheries management, just as it makes no sense to have fishers involved in farm management. As such, members of the village have no place in the management of the fishery.

Members of the resident community, however, had strong counter-arguments to the latter views:

(a) Outsiders cannot manage the lake because eventually they will go back to their original homes, rendering management efforts ineffective.
(b) The resident communities should be the ones to manage the lake because they will always be there.
(c) The fishers continually demand and claim ownership of land traditionally used by the villagers, who see giving occupancy to fishers as undermining their claims over the ownership of the land.
(d) Migrants should abide by the norms and customs of the local residents.

Resident elders at Kabangaja did not believe that immigrants to their village could be fully committed to the management of the lake's resources. One commented that '...the fishers are looking for money. How can they manage this lake?' (Lukona, pers. com.) The fishers, however, retorted that they migrated as they searched for a good catch, good prices, incomes, enough drying land and other opportunities. Anything to do with management was therefore necessary for their livelihoods.

Local residents at Kabangaja recognised they did not fully 'own' the local institutions responsible for the implementation of fisheries management. They queried the reason why the village government could not have fisheries management roles. Instead, a whole, separate institution was formed for this purpose within a day. Hamid (1997: 159) points out that the creation of new organisational structures without taking into account the social processes behind existing institutions might lead to intra-community conflict by dividing the local population or by causing the legitimacy of existing institutions to be questioned.

Cultural and norm variation from this case study reveals difficulties for enhancing the participatory management of common resources. In discussions with local residents, they agreed that the relationship that they shared with the immigrant fishing community was controversial. When the immigrants had first arrived in 1989, the resident community claimed that they had explained much of their culture and preferences to the new arrivals. In 1992, more immigrants arrived under the pretext of being there on short visits as 'uncles', 'brothers' and other relatives.

In Sukuma culture, residents explained, one who came to settle was respected and valued. One respondent explained that their sons were trained to build their own houses in the family compound before they married. This settlement, it was said, would encourage him to develop the land as a livelihood within his compound and neighbourhood environment. 'When I want my daughter to get married, I will advise her to seek a settled man, not migrants, and I would caution her to worry about cultural differences and behaviour which might not be the same as ours and hence lead to an unstable marriage' (Ngika, pers. comm.). Being a permanent settler implied a commitment to the interests of the community as a whole. The group distrusted immigrants with a culture and interests different from theirs.

Fishers at Kabangaja were also asked who they thought should decide 'when fishing should take place'. Such restrictions were interpreted in three ways: first, in terms of normal closed seasons that affect the fishery (between January and June); second, in terms of seasons when high catch values can be expected, and others when values are expected to be low; and, finally, in terms of local beach restrictions that demand that all fishers leave and return at the same time, so that theft can be better monitored. 23% of respondents said both the FD and the community together should decide when fishing could occur. 30% said that the (fishing) community alone should have the responsibility; and 47% (N=14) felt that the FD alone should determine when fishers could leave the landing to fish.

The conflict between the immigrant fishers and resident community at Kabangaja has even manifested itself in matters such as the weather. *Dagaa* fishers prefer sunny and dry weather in which to dry their *dagaa*. When the rains then fail, community members have been known to blame the fishers for using witchcraft to hold back to rain. It was within this context that the fishers argued that if the resident community were allowed to decide when fishing should be allowed, they might close fishing during the dry season, a time when *dagaa* fishers enjoy greater profits. Wilson *et al.* (1999: 566) discuss a case in which a *dagaa* investor was accused of using witchcraft to prologue a dry spell, while Nile perch fishers suffered. At Kabangaja, villagers also thought that there was merit in a closed season during the rainy season, so that the fishers could participate in farming. The fishers, however, retorted that the villagers should already know that the lake is their farm. As a result of these difficulties, many respondents argued that the FD should have responsibility for this decision because conflict could then be avoided.

Members of the community argued that they should be allowed to determine when fishing should take place because as long as they were the permanent residents at the landing, they remained the main beneficiaries of the lake. They also added that it was important for them to make decisions concerning the disciplining of the youth, and preventing drunkards loitering around during the farming season. They said that during the dry season they had no problem with people being fully involved in the fishery or elsewhere. All of the village's labour was needed during the farming season. For this reason, leaders at another survey beach, Mwasonge, had in the past banned festivities and brewing during the farming season (Medard *et al.* 2000). Kabangaja residents echoed these sentiments.

35% of respondents felt that the FD should be the only institution with the power to decide who could or could not fish. Respondents in this group argued that if the FD was responsible for this decision, then the chances of conflict between fishers and the community would be minimised. 39% of respondents felt that both the FD and the community should have this responsibility, while 26% thought that Kabangaja's fishing community alone should have this right. Amongst the respondents in the latter groups were those who argued that if the FD was to have this responsibility, then they would favour only Nile perch fishers.

Fighting over fishing grounds between Nile perch and *dagaa* fishers has also occurred as a result of the former's intense drive to secure adequate Nile perch supplies. *Dagaa* fishers explained how Nile perch fishers forced them to set their lamps in less productive shallow waters. Sometimes, *dagaa* fishers could be prevented from fishing in productive fishing grounds, have their nets stolen, and be chased away (cf. Medard, 2000b).

When asked who should determine where on the lake fishing could take place, 41% of respondents answered that the fishing community should have this right because only they knew where the *dagaa* was. 31% felt that it should be the FD that decided, while the remainder said that it should be the community and the FD together (N=29).

Fishers argued that when managing the fishery in the future, it was necessary for education to be given so that people could learn about the negative effects of illegal gear. They said that some beach seine fishers did not believe that their fishing technique was destructive. In addition, irrespective of what the government had in store for them, they were the key informants on the ground, and not farmers. They complained that the resident community was intransigent, and that they should be more willing to embrace a mixed community.

The resident community said that they could work harmoniously to manage the fisheries resources with *dagaa* fishers and the FD, provided they all worked together to analyse and understand their situation and mutual experience. They would not, however, tolerate being told what to do by outsiders. They insisted that the migrant fishers would have to change to abide by local institutions, culture and perspectives. Said one elder, '...initially, we had genuinely thought that the migrant fishers had our best interests at heart. We did not worry about it then, but we do now because of their negative attitude' (Lukona, pers. comm.).

Additional sources of conflict at Kabangaja community

Other socio-cultural problems have deepened the conflict between the migrant fishers and the resident community. Amongst these has been the question of land ownership. As the community of migrant fishers grew, their demand to be allowed to erect both temporary and permanent structures increased. They also wanted land on which to bury their dead. Members of the resident community, however, started to claim that their farm land extended right up to the lake's edge, and whoever constructed anything upon it would have to pay a fee.

The fishers refused to pay, saying that land up 60 metres from the shore belonged to the government. They then asked the FD to intervene in the dispute. In response, the FD at Ward level issued a series of written instructions:

(a) All fishers should have fishing licence.

(b) They must use the area for fishing only and should not construct permanent structures.

(c) Once they movc away, the land they have been using should be surrendered to the responsible FD office.

(d) It is an offence for anyone to ask for payment for the land assigned to someone by the FD.

(e) It is an offence for community members to humiliate anyone, and to create social unrest amongst themselves.

The fishers reported that this intervention had helped matters, although they still had to buy land to bury their dead. They managed to get a plot some 5 km from Kabangaja, for which they paid Tshs. 170,000/- (US$ 212). Those who wanted to build permanent homes managed to get land elsewhere.

Despite the FD's intervention, resident community members were still concerned about a number of issues. They said that in the old days, the traditional chief, the '*mtemi*', and his land committee arranged land ownership along the lake shore. As far as they were concerned, the community had owned the land, from which they bathed and collected water. The community had only learned that the state owned a 60 metre band of lakeshore land when Mwanza had become a municipality. They wondered why it was that wealthy folk were building businesses and homes within the band, and claimed that rich land owners had been selling land that reached right up to the water's edge. They felt that the whole matter was very contradictory, and, in any case, were concerned about the FD playing a land planning role.

As mentioned earlier, most of the *dagaa* fishers are Muslims. According to Muslim practice, when an adult dies, only men will attend the burial ceremony. Women will remain behind at the compound of the deceased, helping in the kitchen. Resident community members, however, took offence at this, and claimed that when the Sukuma bury their dead, then the whole community is supposed to turn out, and women are not supposed to be excluded.

The village residents explained that at one time a two day old baby from the fishing community had died. Like many other African ethnic groups, the Sukuma do not believe that the burial of one so small needs to be attended by the whole community. Instead, only a few elderly women performed the burial without mourning. This meant that the usual financial contribution to the home of the deceased was not made, and the fishing community was gravely offended by this, and decided that they would not, in consequence, attend the funerals of resident community members. The resident community then took the decision that no land would be offered for Muslim burials, and threatened any one of their members with fines should they attend a fisher's burial.

The deep mistrust that this conflict generated permeated all aspects of the relationship between the resident community and the immigrant fishers. For example, when an international NGO offered to sink wells for the whole of Kabangaja, they proposed that each member of the community contributed Tshs. 200/- (US$ 0.25) towards a maintenance fund. The fishers did not believe that all members of the resident community had made their contributions, while the resident community carried out thorough follow-ups on all fishers to make sure that all had paid. The fishers interpreted the latter action as unfair.

The implications of conflict for the management of the lake's fisheries

Respondents at Kabangaja claimed that their new Beach Management Unit (BMU) was ineffective because of the conflict between the fishers and the resident community. The new BMU confirmed that apart from achievements in beach sanitation, it has been ineffective insofar as fisheries management was concerned. For example, whenever the BMU sought to make licensing follow ups, the fishers claimed not to recognise it, and told its officers that they would only accept orders from the Fisheries Department (FD). Herein lies the problem of inadequate representation of *dagaa* fishers in the new BMU.

It is clear that the perspectives of the two communities involved in this study are related to their preferences. The differences in their knowledge about the management options available to them, their perceptions of likelihood of success, and their relationship as opponents are forcing a wedge between them. In addition, the groups are unequal, and have varying degrees of access to political power, varied ethnicity, social power, class, influence and other factors, all of which define and affect the avenues open to the groups. Income vulnerability, levels of education, poverty, and labour patterns may also affect the ability of people to act on disputes. The absence of clearly defined legal procedures is an additional difficulty.

The fishers complained that the FD had recognised the new BMU and even provided them with equipment, without ever having satisfactorily explained why the original BMU was disbanded. There is a need for a common understanding of what had happened after the formation of the first BMU, what caused the change and who initiated it.

Barrow and Murphree (1996: 4) suggest that cohesion is a necessary facet in the development of common identities and interests that serve to bring people together for collaborative action. Its source commonly arises from a shared history and culture, although it my be a product of political and economic factors which can force people to share a finite resource base (Barrow and Murphree, 1996: 4). If fisheries management is to succeed, it is necessary that an interactive process is developed, organised in such a way that communities can easily adapt to its demands over time.

The disputes at Kabangaja that concern the management of the fisheries resource base are often tangled, complicated and derived from long-standing conflicts between individual community members, families, institutions and other social groups. It is important to note that a seemingly minor dispute may have major implications due to the socio-economic, political or cultural conflicts in which it is embedded. The task of resolving such conflicts can be time consuming, costly, difficult and even impossible. A key starting point would be to recognise conflicts and to integrate them into national Tanzanian fisheries policy to assist decision makers, administrators and planners to understand the conflicts that exist in different fishing communities and that these ought to be addressed in a participatory manner.

The lack of official awareness regarding conflicts is often the result of the centralised fisheries management system established during the colonial period, in which there is only minimal public involvement. As a result, conflicts are only dealt with superficially. When they arise, they are commonly viewed by officials as errors to be ignored or hidden, rather than as pressing problems to be addressed and learned from. Increased community participation opens doors for democracy and equity. Marks (1991: 353) suggests that in order for this to occur, local participation should be analysed in terms of who participates, what institutions are involved and what functions and objectives they have. Such an investigation is necessary at Kabangaja so that a compromise solution may be reached in the conflict that afflicts the two communities there. It is important to start with appropriate training and the laying down of permanent participatory structures for decision-making. All people should be involved, so as to become more responsible

The mistrust and tension between communities in the fisheries sector are becoming apparent at various landing sites. Fortunately, the Tanzanian fisheries sector increasingly perceives the need to seek local participation in the management of the fishery. This process should go hand in hand with community participation in the policy formulation process. Such a process can best be conceived as an ongoing negotiation amongst all members of society, seeking to reconcile their different interests according to the prevailing situation. As pointed out by Sarin (1996 in Castro 1997: 200), '...any community forestry intervention changing the existing resource use pattern impacts on its different constituent groups'. Again, a key challenge is getting broadly based participation in policy formulation so that people's interests can be covered.

Concluding thoughts

The conflicts between the resident village population at Kabangaja and migrant *dagaa* fishers is multi-layered and complex. It initially drew its impetus from ethnic and cultural differences: the village community is mainly of Sukuma ethnicity and Christian, while the immigrants are from the Ha ethnic group and Muslim. These differences developed into complaints that the immigrants failed to respect and emulate local customs. In particular, these complaints have homed in on powerfully emotive subjects such as burial and grieving, and assistance with labour during peak farming times.

It is against this background that the Tanzanian Fisheries Department arrived to thrust upon the community the whole idea of Beach Management Units (BMUs). These drew their membership from community members, and were supposed to hold sway over the implementation and enforcement of government regulations concerning the fishery. In this sense, the introduction of the BMUs represented a mechanism through which access to the fishery could be controlled. In view of the already existing divisions within the community, the controls that the BMUs represented were almost certainly going to augment and sharpen the nature of this conflict.

Indeed, the arrival of the BMU has created a forum through which the community's grievances have now become honed and even, to some extent, multiplied. The question of who owned the land upon which the immigrants have settled is now a source of considerable grievance to the villagers, while a large debate on who holds responsibility for the fishery has commenced.

Kabangaja community members need to develop their management experience with respect to the fishery, not only so that they may solve their dilemmas, but also so that they may develop the necessary knowledge base to manage the institutional, social and economic tasks that they face. In doing so, the managerial process could be adapted to better serve their livelihoods.

Clearly, some conduit for the relief of Kabangaja's conflict is needed, which will enable the communities involved to express and solve their grievances, and where they may develop the commonality of their collective experiences in a way that may contribute to the management of the fisheries resource base.

The Kabangaja community is afflicted with poor lines of communication between the fishing and resident communities, and a lack of trust between the two of them. These problems can only be remedied if all agree to accept their different cultural backgrounds, breaking down stereotypical perceptions of one another, and paving the way to constructive communication.

Solutions to these differences cannot be imposed from outside through, for example, the imposition of BMUs. The characteristics of Kabangaja's conflict are internal to their community. Outside solutions to conflicts so contingent on fluctuating internal socio-political dynamics are unlikely to work.

There is an urgent need to address issues of conflict in national fisheries policy so as to integrate conflict management into fisheries and other resource-based sectors. BMUs were introduced for the purpose of better management, but many did not perceive them in this way. The only way to ensure that such differing perceptions are changed is by raising the level of effective interaction between all interest groups and user groups. This is possible if conflict is transformed into opportunities for change by both immigrant fishers and the native community to finally enable policy makers and communities to formulate applicable laws and regulations for sustainable fisheries management. It is therefore important that co-managerial efforts are designed to encourage and accommodate various lessons. Opportunities must be created for reflection and a critical examination of the successes and weaknesses of various programmes so that adaptations can be made to meet specific demands at specific locations.

Acknowledgements

This paper draws on research carried out at one research site as part of a M. Phil. thesis funded by the EDF-funded Lake Victoria Fisheries Research Project, whose support the authors gratefully acknowledge. The authors are also grateful for the support of the Tanzania Fisheries Research Institute (TAFIRI), and assistance in the field from E. Mlahagwa, B. Zenge, M. Kabati, D. Komba and D. Msunga. Constructive comments from the editors and anonymous referees are acknowledged with thanks. Responsibility for all errors remains with the authors.

11

Institutional Evolution at Lake Chad: Lessons for Fisheries Management

Matthias Krings and Marie-Therese Sarch

Lake Chad is an important wetland in the semi-arid Sahel corridor, which provides the basis of many thousands of livelihoods that depend on its seasonal fluctuations to renew fish stocks as well as to water farmland and rangeland. Despite the millions who rely on fish production from the lake, state management of the lake's fisheries resources is largely ineffective. In the early 1990s, immigrant fishermen from Mali and Hausaland introduced a new fishing technique involving traps, and locally known as '*dumba*' fishing. On the Nigerian shores of the lake, the new technique brought about various conflicts. This chapter outlines how communities in neighbouring areas of the lake shore addressed these conflicts and examines the institutions that were created to resolve them. Despite the promulgation of the 1992 Inland Fisheries Decree by the Federal Government of Nigeria, these institutions were created largely outside the sphere of government influence. Nevertheless, they have had some measure of success in conflict resolution and this chapter considers the insights that their evolution provides for the wider issue of inland fisheries management in Africa. These do not provide optimistic reading for those seeking to promote the co-management of fisheries between local communities and government fisheries departments. Rather, these case studies demonstrate the importance of local elites in determining fishing rights, the extent to which their influence in non-fishing arenas enables them to manipulate interest groups within the fishery and the divergence between their objectives and those of the federal fisheries department.

The evolution of these institutions was investigated through two separate research efforts in neighbouring Local Government Areas (LGAs) of the Nigerian lakeshore, Kukawa and Marte. While the first of these was designed to investigate how the state's fisheries management efforts might be better informed by an understanding of the traditional resource allocation institutions operating on the lake shore, the second, in Marte LGA, is part of a study of community formation within recently established multi-ethnic settlements on the former lake floor[51]. The chapter is structured around the analysis in each

[51] Research was conducted under the auspices of the Joint Research Project between University of Maiduguri and University of Frankfurt, financed by an agency of the German government, the 'Deutsche Forshungsgemeinschaft'.

LGA and concludes with a synthesis of the lessons that can be learned from the common themes in both analyses. Before this, however, the chapter outlines the environmental and socio-political background to the evolution of the fisheries management institutions that are its focus.

Background

The Lake Chad basin covers a large part of central Africa (see Figure 25). The lake itself lies at the south-eastern extreme of the Sahara Desert and traverses the Saharan, Sahel and Sudan-Savannah agro-climatic zones. The lake's catchment area extends into humid zones further south and the Chari/Logone river system, the lake's major influent, drains into the lake from the Central African Republic and the Adamawa highlands to the south.

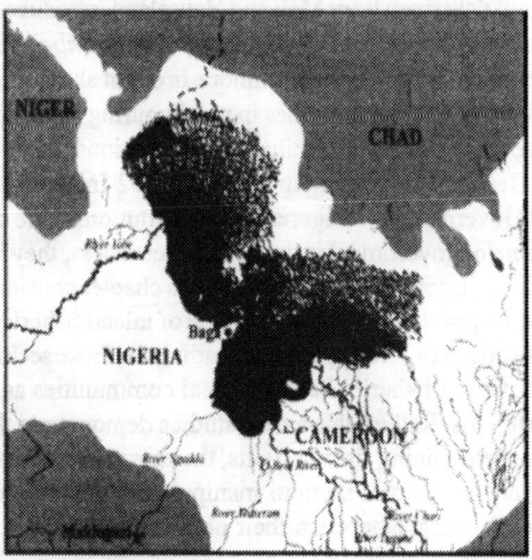

Figure 25: Ecological Zones in Lake Chad during the 1990s

Water from the Chari/Logone rivers flows into the lake at its southern extreme and then flows northwards and outwards encouraged by the lake's gradient and prevailing winds. This inflow peaks in October/November following the end of the rains in the southern catchment area and reaches a minimum in May/June at the start of the next year's rains. The flood waters take between one and two months to reach the study region on the south-west shore where water levels peak in January and reach their minimum in July (Olivry *et al.*, 1996). In the past 25 years, annual rainfall in much of the catchment area has been reduced and the surface area of the lake has varied considerably both on an intra and inter-annual basis (Sarch and Birkett, 2000).

The western shore of Lake Chad has been under the jurisdiction of Borno, one of Nigeria's 36 states, since the end of the fourteenth century. Although the administrative status of Borno itself has varied, it has been dominated by the Kanuri ethnic group for most of its existence (McEvedy, 1995). The Kanuri aristocracy in Borno was named the 'Native Administration' by the British colonists (and is called the 'traditional administration' in this chapter) who collaborated with them to develop their system of taxing the rural population (Temple, 1919). Under the colonial system of taxation, the *Shehu* nominated District Heads, '*Ajia*', who were responsible for collecting tax from the various regions throughout Borno. The *Ajia* delegated this task to sub-district heads, called '*Lawans*', who usually delegated to local agents known as *Bulama*, all of whom were expected to channel revenues upwards to the *Shehu*. Initially, in 1905/6 when this system was set up, the *Shehu* was required to pass half his receipts to the British (Palmer, 1929).

Since Nigerian independence in 1960, a modern government has operated in parallel with the traditional administration and consists of three tiers: Local, State and Federal. There are five Local Government Areas (LGAs) adjacent to the Nigerian shore of Lake Chad. Although LGAs have a fishing and agricultural remit, the level of involvement in fishing and/or farming varies between each LGA. The study region includes Kukawa and Marte LGAs .

The Borno State Government has a minimal involvement in the administration of the lake and its immediate vicinity. This is partly due to international tensions. Armed clashes and rebel activity on islands in the lake have persisted since the 1970s and are largely associated with the succession of civil wars in the Republic of Chad. A multi-national 'Joint Patrol' has been created in response to these outbreaks and has been monitoring the lake and attempting to prevent further violence. Along the western shore of the lake, the Nigerian army dominates the Joint Patrol. The Joint Patrol also includes the mobile police, immigration and customs services.

Other than the agencies involved in the Joint Patrol, the only Federal Government presence in the lakeshore region is that of the Chad Basin Rural Development Authority (CBRDA). The CBRDA made huge investments in irrigating areas of the lake shore in the 1970s. Three irrigation schemes were planned during the peak lake levels of the 1960s and then implemented in the 1970s, when the lake levels reached to lowest levels of the century. These included the South Chad Irrigation Project (SCIP) in Marte LGA. The SCIP involved the construction of a 30 km intake channel designed to pump water from Lake Chad's southern basin to 100,000 hectares of clay plains to the south-west of the lake for rice and dry-season wheat cultivation. Since the implementation of SCIP in 1974, the lake level has rarely been high enough in the rainy season to reach the intake canal. Rice cultivation has been a rarity and wheat yields have been poor (Kolawole, 1986; 1987a; 1987b).

Despite the importance of the contribution of inland fisheries to Nigeria's fish consumption (FAO, 1996b), the management of the nation's inland fisheries is a relatively recent concern of the federal government. No national legislation regarding the licensing or regulation of the inland fisheries of Nigeria was enacted until the Inland Fisheries Decree of 1992. The decree aims to protect inland fisheries resources and enhance their productivity (cf. Ita, 1985: 169). It charges the Commissioner for Agriculture in each state with the responsibility for licensing and regulating inland fishing. Certain regulations on gear are introduced in the decree and there is provision for the creation of further regulations at federal level. Importantly, the 1992 Decree also restricts the damming of inland water[52] and, in effect, prohibited *dumbas*, the catalyst of the institutions that are the subject of this paper. The development of the fishery on the lake-shore is considered next.

Fishing has played a significant role in livelihoods throughout the lake basin for centuries. In the past, the Yedina ethnic group dominated the fishery in the interior of the lake, while the Kanuri, Kanembu and Shuwa Arab groups fished the immediate offshore region. None of these groups relied solely on fishing to make a living. The Yedina and Shuwa Arabs were renowned for livestock rearing (and their fishing) while the Kanuri and the Kanembu were renowned for farming. However, other groups of full-time Kotoko fishermen from the lower Shari river frequently undertook fishing expeditions to the lake. This chapter focuses on the western shores of the lake where the Kanuri have fished for generations.

Until the 1960s, hook lines and gill nets were the fishing gears most commonly used around the lake. Fish fences and traps, though common on the rivers flowing into the lake, were not common on the lake (Couty and Duran, 1968; Blache and Miton, 1962). On the western shores, however, Kanuri fishermen used a technique known as '*tukul*'. *Tukuls* are constructed during the receding flood and consist of dams constructed out of mud and sticks that enclose an artificial pond. The water in the pond is left to stagnate and after two or three days, fresh water is let into the pond. The fish in the pond are attracted by the fresh water and pass through the opening where they are trapped. This technique was developed to provide fish for community subsistence and was operated under the supervision of titled 'chiefs of the water', '*kacalla*'. This office dates back at least to the turn of the century, Since then, however, other titles have been used to describe similar roles. For

---

[52] 'The appropriate authority shall regulate and control the building of dams, weirs or other fixed barriers or obstruction to ensure the free movement of fish, and where permission is granted to a person to build a dam, weir or other fixed barrier or obstruction, fish ladders shall be built to ensure free movement of fish.' (Federal Government of Nigeria, 1992: 1).

example, the title '*kacalla njibe*', 'chief of the water' is currently in use in Marte LGA and '*kaigama*' is used around Baga in Kukawa LGA.

The power of the Kanuri chiefs of the water to grant access to *tukul* sites gradually diminished towards the open water. Besides *tukul* fishing sites, access to the water was not restricted, although chiefs of the water anticipated occasional gifts in cash or kind. This situation has changed since the late 1960s, when the lake attracted many people from all over Nigeria, Niger, Chad and Mali, in particular professional Hausa fishermen from the riverine areas of Kebbi/Sokoto and Hadejia. This influx was stimulated by a combination of several factors, namely improved infrastructure, a growing market for freshwater fish, the introduction of nylon lines, nets, fishhooks, etc. (Sikes, 1973; Van der Meeren, 1980) and the far higher catches of the lake compared with those of the drying up rivers in the immigrants' home areas.

Hausa fishermen introduced new types of gear, namely traps called '*gura*' and '*unduruttu*' (Harris, 1942: 27; Holden, 1961: 154-156). Fishermen began building fish fences by placing small *unduruttu* traps in a continuous line. As water levels declined during the 1980s, new opportunities were opened to catch fish with traps in shallow water. Fishermen from Mali introduced a new type of trap, the '*gurar Mali*', or 'Mali trap'. These were constructed from wooden frames covered with nylon netting. Their heights varied depending on the depth of the water. The traps are placed on the lake floor and fish, *Tilapia* spp. and *Clarias* spp., swam through one-way entrance holes located at the bottom of the trap. Initially, Mali traps were baited and placed in a loose cluster. Early in the 1990s, however, fishermen began to place them in a continuous line, similar to *unduruttu* trap lines, effectively damming water channels emptying into the main body of the lake. These Mali trap lines are called '*dumba*'.[53] *Dumba* fishing has proved so effective that many Hausa fishermen have adopted it.

Big *dumbas* comprise over 1,000 traps and cover a distance of more than a kilometre. They are organised by a headman called the '*uban dumba*', 'the father of the *dumba*', who deals with authorities, pays fees and co-ordinates the work. Individual fishermen who join a *dumba* with their own traps have to pay the headman in cash or kind (Krings, 2000). The traps remain inside the water for several months, during which time, they are emptied every second or third day.

---

[53] Though Kanuri fishermen are reluctant to take part in this type of fishing, *dumba*, or rather, *d*mba*, is a term of Kanuri origin, describing an artificial earthen dam. In the Lake Chad area, such small dams are used in years of exceptionally high tide to prevent the lake water from entering villages. The widespread use of the term '*dumba*' meaning 'fish fence' is probably rooted in the similarity of the two actions.

The dramatic increase in *dumba* fishing soon led to multiple conflicts. Kanuri chiefs of the water complained about decreasing catches, while hook line fishermen - both native and immigrant - were driven away from anywhere near *dumba* sites. Considerable competition for suitable sites in which to locate *dumba* also caused conflicts amongst *dumba* organisers. Traditional administrators were often called in to resolve *dumba* disputes. In 1994, Kukawa and Marte LGAs attempted to license and tax *dumba* and confusion developed over whether LGAs actually had this right. Furthermore, although the Local Governments on the lakeshores do endeavour to play an active role in regulating and taxing fishing in their areas, compliance with measures such as licence fees is limited by the lack of LGA resources and the inability of their staff to reach the most productive fishing areas on the lake and enforce them. Early in 1995, federal fisheries officers visited Marte LGA and explained that the regulations of the 1992 decree prohibited *dumba* and therefore, in theory, prevented LGAs from taxing them. Although news of this prohibition travelled quickly along the lake shore, the use of *dumba* persisted and a range of institutions soon developed to meet the demand for *dumba* allocation systems. The remainder of this chapter examines the evolution of such institutions in Marte and Kukawa.

Access to fishing in Kukawa Local Government Area

Kukawa LGA is situated to the north of Marte LGA on the south-western shore of Lake Chad (see Figure 26).

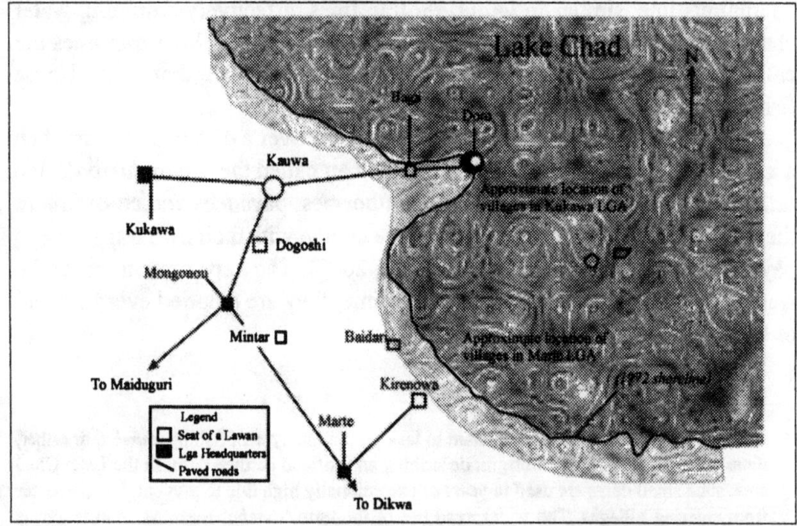

Figure 26: Sketch map of study regions on the Nigerian shore of Lake Chad.

The local government headquarters and its chief executive (who is known either as the sole administrator when appointed by the army or as the chairman when elected) are based in the small town of Kukawa. The main fish market along the Nigerian shore of the lake is Doro, lying to the east of the settlement of Baga. Kukawa is linked by road to Maiduguri, the state capital, and is an important node in the transport of the lake's fish to the large urban markets in southern Nigeria. Baga is also the seat of the traditional 'sub-district head', the *Lawan* under whose jurisdiction many (but not all) of the fishing villages in Kukawa LGA fall. The investigation into fishing rights was based on fieldwork in these and other fishing villages around Kukawa. This involved a series of village head interviews and focus group discussions with village elders in nine villages in 1993 and 1994. These were followed up with participatory village appraisals conducted over two weeks in two case-study villages during 1995. The appraisals were designed to understand the institutional channels of resource access, their context and evolution, and to enable contrasts and comparisons between villages to be made. The further stages of the investigation used predominantly secondary sources to examine access institutions at the district, regional, and national levels. These exercises were undertaken as part of the British Government Fisheries Research Project (Neiland and Sarch, 1993). Subsequent analysis of this information has been undertaken in the UK (Sarch, 1999).

Conflicts in Kukawa

By the end of the 1992/1993 fishing season, *dumba* conflicts had divided the fishing communities around Baga along ethnic lines. The predominantly Kanuri village heads, '*Bulamas*' were opposed to *dumba* fishing and had passed several complaints about the decreasing catches of Kanuri fishermen to both the *Lawan* and the police in Baga. Meanwhile the *dumba* fishermen organised themselves into the Lake Chad Fishermen's Welfare Organisation. The LCFWO was established with the backing of a wealthy Hausa entrepreneur, Alhaji 'Mai Tractor', who operated from Doro. Fishermen were invited to pay a subscription for membership and, in return, the Organisation endeavoured to use these funds to pursue the fishing 'rights' of its members with the police in Baga.

Amongst *dumba* fishermen there was considerable competition for suitable sites in which to locate *dumba*. This led to disputes and occasionally violent conflict. In 1993, such conflicts in the villages around Baga were resolved by the *Lawan* of Baga, who agreed to licence the use of *dumbas* at particular sites in his jurisdiction. These licences could then be checked and the exclusive rights of the licensee enforced by the Joint Patrol. Focus group discussions with fishing community leaders during the 1993/1994 season indicated that in most villages visited around Baga, this system was working smoothly. The presumption was that Alhaji Mai Tractor had participated in resolving *dumba* conflicts with the *Lawan* of Baga, and had recommended the new *dumba* allocation system to LCFWO members.

Multiple claims

Claims for rights to exploit *dumbas* were not, however, limited to the Kanuri and Hausa fishing communities of the lake shore. The *dumba* conflicts and the subsequent licence fees soon brought the profits to be made from *dumba* fishing to the attention of the lakeshore LGAs. In the following 1993/1994 fishing season, Kukawa Local Government instructed the village heads, the *Bulamas*, to sell *dumba* licences on their behalf and to pass the revenues to the LGA. Compliance with this directive was limited, and in only two of the nine villages visited in 1994 did the *Bulama*s issue LGA *dumba* permits. In the remaining seven, *dumba* revenues were passed on to the *Lawan* of Baga. The two villages whose *Bulama*s had complied with the Kukawa LGA directive were under the jurisdiction of the neighbouring *Lawan* of Dogoshi (see Figure 26). The resulting confusion over which authority had the right to license and tax *dumba* was resolved at the beginning of the next fishing season, in early 1995, with the visit of the federal fisheries officers to enforce the regulations of the 1992 decree. As these prohibited *dumba,* they prevented LGAs from licensing and taxing them and the Kukawa LGA was obliged to stop enforcing its *dumba* regulation initiative.

Powerful individuals

Following the withdrawal of the Kukawa LGA from *dumba* licensing and taxation, the *Lawan* of Baga, together with neighbouring traditional administrators, rapidly exerted his authority over the allocation of *dumbas*. The ability of the *Lawan* of Baga to do this reflects the power and influence that he wielded over the communities under his jurisdiction. Although *dumba* fishing was not a traditional source of revenue for the *Lawan*, farming was. In most of the villages established following the recession of the lake in the 1970s and 1980s, the *Lawan* had requested the village community to nominate a *Bulama* through whom they should channel their annual taxes. Although taxes are never welcome, they were not unexpected, since a broadly similar system, in which local rulers allocate and tax land, has persisted throughout northern Nigeria (see Hill, 1972). In return the *Bulama* was given the *Lawan*'s authority to allocate residential and farm land, and to settle disputes within his community. Thus many of the communities on the Kukawa shores of the lake are obliged both to pay taxes to the *Lawan* and to recognise his authority.

Resulting institutions

Although access to fishing grounds varied between villages in the Kukawa LGA (depending on both the season and location of the fishing ground), in general, exclusive rights to fishing from a *dumba* site were sold for cash, in advance, on a seasonal basis. In the villages around Baga, fishermen seeking a *dumba* licence were referred to the *Lawan*'s representative, the '*Kaigama*' (a traditional post historically allocated to senior slaves by their overlords). Village-based agents of the *Lawan*, either the *Bulama*, *Wakili* or assistants, acted as a conduit to the *Kaigama* in Baga. After negotiation, the *Kaigama* issues a licence, a note containing a description of the *dumba* location and the dates between which the *dumba* is permitted. This is stamped with the words 'Native Water Controller, Baga-Kauwa' and dated. In the 1994/1995 season, exclusive rights to operate a *dumba* at a particular site were sold for as much as 10,000 Naira, (over US$100) in 1995. The *dumba* permit is then taken to the local headquarters of the Joint Patrol, where the licensee negotiates a fee for the army to endorse the licence. For fishermen less able to pay the large advance fees charged for exclusive access to particular *dumba* sites, another option is to buy a temporary concession from the licensee of a particular site. Here, the licence holder must be found and negotiated with before fishing can begin. The licence holder usually expects a proportion of the catch, or its proceeds, as a further payment.

Whatever the circumstances of *dumba* acquisition, recent knowledge of the fishing industry is crucial. Knowledge of whom to pay and how much is feasible is important and the system favours fishermen with good contacts with the traditional administration. The importance of such good relations is illustrated by the Hausa fishing community of Dabar Shatta Kwatta, which attempted to acquire community fishing rights to a *dumba* site in the 1994/1995 season. Although they had paid and acquired a licence stamped by both the 'Native Water Controller' and the Joint Patrol in Baga, a group of Kanuri fishermen exploiting the site ignored their licence and the fishermen of Dabar Shatta Kwatta were unable to obtain any redress from either the Joint Patrol or the *kaigama*.

Although the rights to allocate access to the fisheries of Kukawa LGA were claimed by a range of organisations, the 'modern' post-independence agencies that have been established to develop fishing resources and livelihoods had have little influence on access to fishing. Although the Joint Patrol, a federal government agency, does play a role in allocating exclusive access to fishing grounds, this is outside their remit. In contrast, the customary institutions of the 'traditional' administration are much more important in explaining access to fishing grounds. The primary objective of these 'customary' institutions is profit. The revenue collected from the taxes is not re-invested in the lake shore and the 'traditional' recipients of tax revenues do not account for them.

Successful institutions?

At the end of this investigation (1995/1996 fishing season), *dumba* sites throughout Kukawa were being allocated by the 'traditional' administration so as to collect revenue and prevent conflict. The second of these objectives is shared with the Joint Patrol whose officers also profit from their endorsement of the *dumba* licences issued by the 'Native Water Controller', such as the *Lawan* of Baga. The profits to be made from *dumba* fishing were reflected in the ubiquitously high licence fees that are charged for them.

Unlike the case of farming land, there is no widely established system of regulating access to fishing in northern Nigeria. The system described above was a recent creation that developed in response to the introduction of the highly profitable *dumba* fishing method that had led to conflict and confusion. Conflict between fishermen developed over the impact of *dumba* on downstream fisheries and confusion existed between modern administrative agencies over *dumba* regulation and taxation. Large *dumba* profits were both a source of conflict and the motive for the traditional administration to resolve the conflict and benefit from the profits. During the remainder of the investigation (up until the 1995/6 season), the *dumba* allocation system operated by the *Lawan* of Baga was largely successful both in preventing conflict and earning significant revenues from *dumba* fishing. Disputes did arise, such as that in Dabar Shatta Kwatta, but these did not escalate to the violent conflicts that preceded the *Lawan*'s intervention.

The absence of a widely used system for allocating access to fishing meant that potential *dumba* fishermen needed to not only find the right person to negotiate with, but also to find out how the system of access was operating at a particular location and time. Although not successful for the fishermen of Daba Shatta Kwatta, the system established by the *Lawan* of Baga also succeeded in reducing the transaction costs for the fishermen who were able to use the system to establish exclusive access to *dumba* sites.

The success of this system in raising revenues, preventing conflict and in particular, reducing the transaction costs of *dumba* fishermen was largely dependent on the confidence which those using it had in the authority of the *Lawan* of Baga. This was curtailed abruptly following his death in 1998 and the subsequent succession process.

Access to *dumba* fishing rights in Marte LGA

Marte is situated to the south of Kukawa LGA on the south-west shore of Lake Chad. The LGA centres on the town of Marte that is the seat of both the traditional district head, the *Ajia* and the Local Government Chairman and his Secretariat. Marte LGA is also the location of the Chad Basin Rural Development Authority's South Chad Irrigation Project (SCIP). The town of

Kirenowa is the location of the pumping station designed to deliver lake water from an intake channel to the clay plains of the lake shore. Like Kukawa, Marte LGA is linked by paved road to the state capital, Maiduguri.

Unlike in Kukawa LGA, the *dumba* fishermen of Marte LGA were successful in establishing an association with which they controlled access to the water and the allocation of *dumba* sites. The following analysis reconstructs the evolution and history of the Fishermen Association Marte Local Government (FAMLG) with a special focus on the role of traditional and modern powerbrokers from within and without the fishing industry. The analysis is based on about 10 months of fieldwork from 1997 to 2000. Data were collected through participant observation, informal discussions and tape recorded interviews with ordinary fishermen, functionaries of the FAMLG as well as traditional and modern officeholders.

Conflict in Marte LGA

In the season of 1993/94, Hausa fishermen began experimenting with Mali traps at the lake margins of Bula Butube (near Kirenowa). Local Kanuri fishermen strongly opposed the immigrant fishermen and - under the leadership of one chief of the water – began to remove and destroy their traps. While the Hausa saw envy as the motivating force of their opponents, indigenous inhabitants were shocked by the efficiency of the traps and consequently feared a drastic decline in catch. The Hausa took their complaints both to the traditional District head, the *Ajia*, and to the Local Government headquarters where the soon to be chairman had an open ear for them. He asked them to pay revenue for their traps and cautioned the Kanuri to work together with the immigrants. When, in the following season, inspectors of the Federal Department of Fisheries arrived to enforce the ban on *dumba* fishing, Kanuri fishermen saw their chance for revenge. Despite the attempts of the Local Government chairman to persuade the inspectors that there were no *dumba* sites in Marte LGA, the inspectors went out into the lake and, with the aid of local Kanuri fishermen, the inspectors discovered several *dumbas* and destroyed them..

The Hausa fishermen, whose gear had been destroyed, began to collect large sums of money from one another with the intention of taking the destroyers of their property to court. The money was intended to be used as a bribe to win the case. The chairman is said to have backed their interests. The case, however, never reached a court since the *dumba* fishermen later expressed their desire to be allowed to continue *dumba* fishing rather than obtaining financial compensation for their destroyed gear. Eventually, sums of money were exchanged and it was agreed that, officially, there were no *dumba* fences in Marte LGA. The Federal Department of Fisheries never returned to undertake an inspection.

Multiple claims

Following the relinquishment of the Federal Fisheries Department's claim to prohibit *dumba* fishing, the LGA chairman proposed the foundation of an association which would enable the remaining interest groups in the fishery to mediate their claims to the fisheries of Marte LGA. The proposal was for an association directed by a multi-ethnic committee of fishermen who would allocate rights of access to *dumba* sites. The fisheries supervisor of Marte LGA registered the association at Maiduguri with the Borno State Ministry of Youth, Sports and Development in 1995.

Powerful individuals

The support of the Marte LGA chairman was essential in the establishment of the Fishermen Association in 1995. Besides his political office, he derived power from his descent from a locally well-known Shuwa family, which provided him with good contacts with both traditional administrators and the 'modern' politicians at state and federal levels. A system of mutual aid or reciprocity soon developed between the Association and the chairman. He protected the interests of *dumba* fishermen and let them govern their affairs more or less independently. In return, the Association assisted him with revenue collection and, perhaps more importantly, by successfully campaigning for him during the Local Government elections in 1997 and 1998.

Resulting institutions

The Fishermen Association Marte Local Government (FAMLG) has a structure similar to that of political parties, unions, self-help groups or social clubs. This structure is characterised by a hierarchy of posts ranging from chairman, vice-chairman, secretary, auditor and others to ordinary membership. The FAMLG has a written constitution listing all members of the committee. It issues laminated membership cards, and is formally recognised by the Local Government authorities. The posts of the committee were shared amongst the three major ethnic groups of the region: the Kanuri, Shuwa and Hausa. The honour of grand patronage was given to the chairman of the Local Government. The same Kanuri Chief of the Water who had formerly led the opposition against *dumba* fishing became the Association's chairman. The vice-chairman was a Hausa fisherman who, due to his charismatic personality, had been a prominent member of the Hausa fishermen's community for a long time. The post of secretary was held by a Shuwa Arab who was not a fisherman, but able to read and write Hausa in roman script and fluent in Arabic, Kanuri and Hausa. The other committee members were either Kanuri Chiefs of the Water or prominent Hausa *dumba* fishermen. *Dumba* heads of Malian origin were only granted ordinary membership.

With the invention of the FAMLG, the taxes or fees that had to be paid by the individual organisers of *dumba* fences were fixed at 25,000 Naira (US$ 250), comprising 10,000 Naira collected by the Local Government as revenue, 5,000 Naira paid to the traditional Kanuri Chief of the Water in whose territory the *dumba* is fixed, 5,000 Naira paid to the FAMLG and another 5,000 Naira shared between the village head (the *Lawan*) and the ward head (*Bulama*), in whose territory the fish fence is constructed. In the 1998/99 season of the receding flood, 15 *dumba* fences were constructed, along with a larger number of smaller fences, for which only half the fee is paid.

The FAMLG was created to represent the interests of the LGA's fishermen, irrespective of their gear. Due, however, to the interests of its leading committee members, the Association's concerns were soon restricted mainly to the management of *dumba* fishing. Rules and regulations for *dumba* fishing were developed, the most important of which concerned the location of *dumba* fences. New rules determined that neither *dumba* nor *tukuls* could be built in such a way that other *dumba* or *tukuls* were excluded from catches. Other rules protected *dumba* sites from fishermen using hook lines. Finally, rules for the internal organisation of the trap-fence teams were developed, defining the rights and obligations of *dumba* heads and participating fishermen. Violation of these rules were said to be punished by either destruction or confiscation of gear or confiscation of catch.

Site allocation procedures were rather obscure. The former vice-chairman explained that the committee met at the beginning of each season and decided upon the distribution of sites. In most cases, *dumba* owners who had used a site for a couple of seasons, maintained the right to these sites. Until 1999, when the committee was restructured, most of these *dumba* heads were, in any case, members of the committee. Those who intended to seek new sites depended on the bargaining power of patrons within the committee. Every season, rumours spread concerning the exorbitant sums of money, over and above fixed fees and revenue, spent by *dumba* organisers to secure new *dumba* sites. Whether true or not, the rumours at least symbolise the potential financial power of *dumba* organisers who serve as role models for younger or poorer fishermen.

The success of the institutions

During the first three seasons after the establishment of the association, no major conflicts took place. According to a local government fisheries supervisor, the Hausa vice-chairman of the Association, who took care of the Association's day-to-day activities, was known for his rigidity. Sometimes, he over-stepped his authority by punishing individuals himself instead of involving governmental officials. In most cases, a simple threat to call upon the police and other government agencies often ensured that conflicting parties accepted the judgement of the Association. In other cases, the Association confiscated the boats of those reluctant to pay their revenue.

The patronage of the chairman also affected the power structure within the Association. Although nominally multi-ethnic, the Hausa *dumba* fishermen of the committee, led by the vice-chairman and secretary (the only Shuwa Arab of the Association's committee), soon gained control over the Association's affairs. The traditional Kanuri chiefs of the water (*kacalla*), some of who were also committee members, therefore no longer held any power on the committee. While they where formerly able to tax immigrants, or collect their share of a catch, their traditional *tukul* fishing was now treated and taxed in the same way as *dumba* fishing. In early 1999, six out of seven *kacalla* declared that they accepted the Association only because of its alliance with Local Government Authorities. Most of the Hausa *dumba* heads accepted the system established by the association for the same reason, and because it reduced transaction costs and brought them close to the local source of political power. To them, the Hausa vice-chairman acted as a powerbroker, protecting their interests while, at the same time, demanding his own share of *dumba* benefits.

The first substantial conflict occurred in 1998 and marks the beginning of a crisis that finally led to the restructuring of the Association's committee in the second half of 1999. The conflict concerned a *dumba* at Abbaganaram. Like many other conflicts over resources in this region, it was rooted in the reshaping of village areas as a result of a change in political leadership at the LGA headquarters (see Krings, 1998). In 1993, the Hausa ward head of Abbaganaram, Alhaji Biri, had lost his office to the head of the Shuwa population living in the village. During the years that followed, Alhaji Biri had been able to regain power over half of his former territory, with the other half remaining in the hands of the Shuwa Arabs. Unable to regain all of his former territory, Alhaji Biri prevented his neighbouring Shuwa Arabs from joining the three *dumbas* he was operating in the area.

In 1998 at a meeting of the association in Marte Local Government secretariat, Alhaji Biri questioned the whereabouts of fees the Association had been collecting from the *dumba* heads for the past seasons. It was discovered that the Association's vice-chairman and other members of the committee had invested the money in fishing nets that they had been using at their own *dumba* sites. Since the committee made no attempts to refund the money, Alhaji Biri became the leader of a group of members who opted for the dissolution of the Association.

Later that year, at the beginning of the *dumba* season, the Shuwa Arabs of Abbaganaram demanded that the Association allocate them the fishing rights to one of Alhaji Biri's three *dumba* sites. The vice-chairman accepted their request and consulted the village head. Together, they decided to place one of the *dumbas* under the jurisdiction of the ward head of the Shuwa Arabs. This decision was sent in writing to the LGA headquarters, the District head, the *Ajia*, and to Alhaji Biri. Alhaji Biri was unwilling to accept the proposal and over the following months, involved both traditional and modern authorities, ranging from the Divisional Police Officer of Marte LGA to the *Shehu* of Borno, in his campaign to regain his rights to the *dumba* site. The Shuwa Arabs refused to relinquish their rights to the site and rumours of an imminent clash between Alhaji Biri's followers and the Shuwa Arabs spread around Marte LGA. Reconciliation attempts by the Association failed and Alhaji Biri banned the traps of the Association vice-chairman from his *dumba* sites. The actual conflict was finally resolved by the aid of nature: with the retreat of the water in March all three *dumba* sites fell dry.

Even though the Local Government authorities had backed the actions of the Association, these events weakened the position of the Association's vice-chairman. He began to face growing opposition within the Association due to the misuse of fees and his failure to settle the conflict of Abbaganaram. In 1999, he was convicted of having illegally leased out fishing rights on the CBRDA intake channel. This had touched upon the interests of an elite group of Maiduguri-based businessmen, politicians and administrators whose powers and influences far surpassed the power of the vice-chairman's local patron, the Marte LGA chairman. The vice-chairman's career came to a sudden halt when the LGA chairman withdrew his support and dismissed the whole committee. A new and largely non-fishing committee for the Association was nominated by the LGA chairman. The Kanuri chiefs of the water are no longer members of the Association, although it continues to recognise their rights as traditional officeholders, and they still receive their share of *dumba* fees.

For the sake of peace, a majority of Marte LGA's fishermen accepted the new composition of the Association. Despite rumours of the new vice-chairman's incompetence during the 1999/2000 season and predictions of imminent collapse, the Association was still operating during the next season. In early 2001, the new vice-chairman toured the Local Government's waters with an entourage of police officers, chiefs of the water and LGA representatives. This was an indication that he had been able to re-establish his authority amongst the multiple claimants to *dumba* fishing rights, so consolidating his power and prolonging the survival of the Fishermen Association Marte Local Government.

Synthesis and conclusion

These examples have been compared because they describe two different ways in which institutions were developed to resolve conflicts in access to *dumba* fishing rights in different, although not dissimilar, locations on the Nigerian shore of Lake Chad. Important differences have been observed in the role which traditional administrators were able to play in the allocation of rights to *dumba* fishing sites. In Kukawa, the traditional *Lawan* of Baga established himself at the apex of the system for allocating *dumba* sites, whereas in Marte, the chairman of the modern Local Government Area effectively sidelined the traditional *kacalla*. Nonetheless, both institutional stories share common threads and these provide interesting insights for the fisheries co-management debate.

In both cases, systems of allocating access to *dumba* sites were developed to address the conflicts arising from multiple claims on a valuable resource. In the immediate environment of Lake Chad, it is the value of the fish in external markets rather than the local scarcity of the fish that provided the impetus for the evolution of institutions to allocate access to the resource.

In both cases, these institutions had periods of success in addressing the conflicts that brought rise to them and some success in reducing the transaction costs incurred in acquiring access to fishing resources. These periods of success were not characterised by any sense of democracy in their creation. Rather, they were dependent on the ability of those in control of the organisation allocating access to *dumba* sites to maintain their power base. The ability of such systems to function depended largely on a small number of powerful individuals and the influence they had over the various interest groups in the fishery. While fishermen were one of these groups, and thus had some say in the operation of the institutions for access, theirs was neither the only voice nor the loudest. In contrast, while the fisheries 'managers' appointed by government played a role in the initial conflicts, they had no voice at all in the institutions that were developed to resolve such conflicts.

Efforts to impose federal fisheries management rules failed in both cases as local interests overrode them. Even if there were the political will to manage such resources at a federal, or even state level, in Nigeria, the capacity of government at these levels does not allow for such political will to be exercised. Improving on such capacities to implement fisheries management objectives is likely to be futile in a political environment where national/regional government has little impact on livelihoods in such remote rural locations.

An important conclusion from the experiences described here concerns locally-based access arrangements. Where institutions for allocating access to fishing resources have succeeded, they have been developed locally and usually to serve the interests of local people, although not necessarily local fishermen or even fishermen at all. In both cases, demand from fishermen for access to fishing resources provided an opportunity for local elites to profit from their power to resolve the resulting conflicts. Thus, despite the crucial local dimension to the success which these arrangements have had, they cannot be said to be community-based nor co-management arrangements. These institutions neither consider management objectives such as sustaining the fishery (however it is defined) and nor do they consider the objectives of the fishermen whose livelihoods depend on the resource.

It would seem from these examples in Nigeria, and others in this volume, that a key dimension lacking from the co-management paradigm is the crucial role of local elites in allocating natural resources to resource users. Fisheries co-management assumes that the fisheries managers appointed by government should co-operate with fishermen and representatives of their interests. Such direct links cannot be assumed. Fisheries managers need to understand the roles that local power brokers play in the fishery before attempting to shape the way in which its resources are exploited. If managers were able to meaningfully convince power brokers in local fisheries that sustainability was in their interest as well as in the interest of the fishing livelihoods, devolving responsibility for fisheries management to the local level could be a realistic option.

12

Management and Development of the Gambia River Fisheries: A Case for the Co-Management of Inland Fisheries Resources

Momodou Njie and Heimo Mikkola

The River Gambia, its tributaries and floodplains are important inland fisheries for the country. As the river enters Gambian territory, 680 km from its source in the Futa Djallon highlands in Guinea Conakry, it flows along an east-west axis, dividing the country into two almost equal parts (Figure 27). Within the country, the river flows for 470 km, and passes through all of The Gambia's administrative divisions. The river is a major waterway and tourist attraction. Its floodplains, riverbanks and wetlands are important habitats for wildlife and play an important role in local livelihood strategies, being used for rice and other crop production and salt extraction. There is also tremendous potential for aquaculture (oyster/shrimp) development.

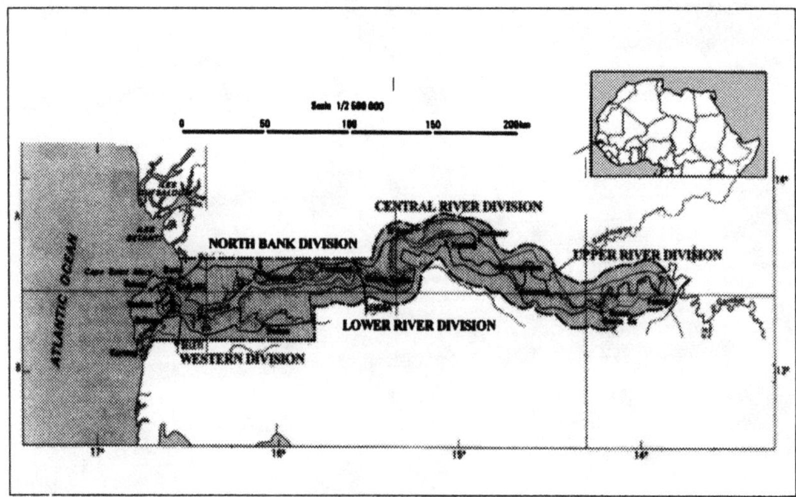

Figure 27: The Gambia:administrative divisions

Above all, the river harbours a diversity of valuable fish stocks in both its estuarine and freshwater parts that provide job opportunities, income and food for thousands of Gambians. The river is fringed on both banks by mangrove vegetation, which provides shelter and breeding/feeding grounds for various marine and freshwater fishes. The estuarine and freshwater boundary of the river moves up and down stream seasonally and, to a lesser extent, tidally. The average boundary lies about 250 km upstream, leaving about 220 km of freshwater in what are known as the Central and Upper River Divisions. The freshwater part of the river contains a number of socially and economically important fish species. Most abundant are tilapia (*Oreochromis niloticus*), African bonytongue (*Heterotis niloticus*), the cyprinid *Labeo senegalensis* and the catfishes *Synodontis gambensis, Clarias gariepinus, Chrysichthus furcayu* and *Auchenoglanis occidentalis*. The lower portion of the River Gambia is brackish and marine fish are common, particularly bonga (*Ethmalosa* spp.), sea catfish, threadfins, barracuda, sole and shrimps. The shrimp (*Penaeus duorarum*) and tonguesole (*Cynoglossus senegalensis*) are purchased by industrial fishing companies for processing and export to Europe (FAO, 1999a).

Between 1977 and 1982, the annual catch from the Gambia River varied from 1,400 to 3,500 t, and the potential annual yield was estimated to be between 2,000 and 8,000 t (Welcomme, 1979). According to the FAO (1999b) inland fish catches from 1984 to 1999 were consistently between 2,700 t and 2,400 t per year. The static nature of these figures indicates that either the fisheries are at maximum yield, or that the statistics are unreliable. What is known for sure is that the presence of large numbers of foreign fishers (from Senegal, Mali, Guinea Conakry and Guinea Bissau) has created intense competition which, coupled with population increases, threatens the sustainable use of the resource.

This chapter reviews the experience and structure of community participation in Gambian fisheries management with specific reference to the management of catfish ('*konokono*') stocks in the country's Central River Division (CRD). It is hoped that such a review will assist in identifying ways in which communities can better participate in the management of the Gambia River and other inland fisheries resources.

Community fisheries centres and co-management in The Gambia

All of The Gambia's fisheries fall under the Fisheries Act and Regulations. Under this Act, the Minister responsible for fisheries is empowered to make regulations for the proper management, development, conservation and protection of fish resources, a particular fishery or species of fish. The Fisheries Department is the technical department responsible for fisheries administration.

It prepares policy guidelines for the management and development of fisheries resources and ensures that regulations are enforced. The Director of Fisheries is mandated to prepare and continually review plans for the management, development and promotion of both inland and marine fisheries resources. Such plans should specify management areas, limits on local fishing operations, and the amount of fishing to be allocated, if any.

The government first concentrated its management efforts on marine fisheries, which are wide open to exploitation by small and large vessels. However, the highly migratory nature of the stocks and the low operating budgets of the Fisheries Department, made regulations difficult to enforce. More recently, with the realisation that inland fisheries contribute favourably towards national policy objectives, and in response to structural adjustment programmes and diversification policies, the Gambian government undertook to develop its artisanal river fisheries. Just as with the marine fisheries, conventional conservation measures, such as mesh size regulations, gear restrictions etc. were applied, but were difficult to enforce because fishing bases are widely scattered and/or hidden and river surveillance patrols are constrained by lack of funds.

The general failure of legislation and enforcement in addressing the ever-increasing complexity and over-exploitation of The Gambia's fisheries resources, led to the search for alternative management models. It seemed logical that the constructive engagement of local communities in decision-making and implementation could help overcome at least some of the problems posed by lack of human and financial resources in the Fisheries Department.

In 1979, the European Union-funded Artisanal Fisheries Development Project (AFDP) established Community Fisheries Centres (CFCs) of which there are now 18: seven along the coast and 11 on the Gambia River. Through the CFCs, infrastructure, facilities and services such as fish landing, handling and processing facilities, mechanical workshops for the repair and maintenance of outboard engines and fishing boats, and community services such as water and fuel supplies, were provided to fishing communities. Fishing communities were organised into trade or user groups and then strengthened with training. Management of the CFCs was the responsibility of a Management Committee (CFCMC), comprising elected representatives of the various trade/user groups, village elders and the '*Alkali*' (village head). Initially assisted by a Fisheries Development Unit (FDU), a multidisciplinary team of fisheries personnel, the CFCMCs were empowered through capacity-building training activities. The management committees then gradually assumed control of the CFCs, with the FDU acting only as an advisor. Their activities included the following:

(a) Coordinating the collection of rental and service charges from users.
(b) Discussion and planning of further development activities.

(c) Repair, maintenance and expansion of project facilities.
(d) Representing the community in negotiations with government and other agencies.

This autonomous management system has developed a relationship between communities and the government based on trust and partnership. The general evolution of the CFC concept, and a detailed description of how CFCs function, are described by N'Jie and Mikkola (2001).

A co-management case study from the Central River Division of The Gambia

The Central River Division (CRD) is the largest of The Gambia's five divisions (Figure 27), covering an area of about 1,500 km^2. In 1993, an estimated 156,000 people lived in the CRD, 15% of the national population. The CRD lies in the middle of the country. Here, the river meanders and forms a network of tributaries that are sheltered by peripheral vegetation dominated by mangroves covering an area of about 67,000 ha. Mangroves are sources of fuel wood and construction timber and provide breeding and nursery grounds for commercial fish species of both river and marine ecosystems. In much of the area, the water is virtually fresh, with salt-water intruding from downstream only during the dry season.

The area is popular for subsistence and small-scale commercial fishing and has become a focal point for small-scale fishers from The Gambia's interior, Senegal, Guinea Conakry, Guinea Bissau and Mali. Gambian fishers have mainly been part-time fishers with farming as their main occupation. Foreign fishers ('strangers') are highly migratory and may fish full-time or part-time. Many foreign fishers have, however, settled permanently in the area with their families.

The influx of foreign fishers into the area intensified during the 1970s and 1980s due to serious drought in the sub-region. There are about 314 fishers (212 Gambians and 102 foreign fishers) scattered amongst 44 recorded fish landing sites or villages in the CRD. Of these, 246 are full-time fishers, while 68 are part-time fishers. Targeted fish species include '*fantangho*' (*Heterotis niloticus*), '*konokono*' (*Clarias* spp.), '*walinyaabaa*' (*Chrysicthys* spp.), '*furu*' (*Tilapia* spp.) and '*suyewo*' (*Gymnarcus niloticus*). Fishing craft are predominantly non-motorised dugout canoes of about 6 metres, with a few bigger, generally motorised, planked craft. The main fishing gears employed in the CRD are the cast net, set nets, hook and line and drift gill net fishing. Hook and line is the dominant gear in the *konokono* fishery.

Konokono is the most important fish species in the CRD. Traditionally, *konokono* was not well liked by the local people, especially fresh, because of its spotted body (associated with leprosy), the soft and slippery texture of its flesh and its flavour. Over time, however, the fish has become well liked by all as a food fish and, more importantly, for the income it fetches when sold. *Konokono* is sun-dried, or occasionally smoke-cured. Prices range from Gambian Dalasis (GD) 10 to 16 (US$ 0.65 - 1.05) per kilogram, depending on the size of the individual fish sold. Much of the fish is exported to Senegal where some US$ 2.00 per kilo can be obtained. According to fishers, the Senegalese then export the fish to France for the African, Latin American and Asian expatriate market where the fish fetch around US$ 5 per kilo. This trade has become increasingly lucrative and more competitive, and many fishers have become rich, at least by local standards.

Konokono is more common in the CRD than in other parts of the river, because its favoured breeding grounds are to be found here. Fishers explain that *konokono* normally breed in the river tributaries and adjacent swamps where they spawn and nurse their young amongst grasses, peripheral vegetation and in ponds and puddles. Spawning normally begins in the early part of the rainy season, when adults move into sheltered river fringes.

Fisheries management in the CRD

Fisheries agents are posted to inland fisheries, where they function as extension agents and are supposed to enforce existing regulations and assist fishers with management and solving problems, including mediation in conflicts. Regulations are pegged to the availability of scientific information about the fisheries. This is usually lacking, and formal regulation is, therefore, limited in its efficacy.

Customary community-based management aimed at the protection and conservation of stocks, and for resolving conflicts, used to be a common feature of Gambian riverine fisheries. It assumed different forms, was very informal and generally weak, and was both direct and indirect. In the CRD, effort controls were employed which combined seasonal closures, fishing gear restrictions on specific species at specific periods, and closed areas imposed to protect breeding populations.

Customary law in the CRD's agricultural and land sectors dictated that families could claim ownership to land passed down to them by their ancestors. Conversely, the river's fisheries resources were generally considered open access and such customary laws did not apply to it. Communities, however, used to lay claim to ponds and pools located in or near their rice fields. Only the families to whom the land belonged could exploit these. Owners used traps, baskets, mats etc. to catch fish. The whole community also had access to community-owned ponds located on the floodplain of the river. The harvesting of these pools was done mainly at the end of the rainy season when the floodwaters receded and fish concentrated in pools. At this time, before the drought period, community elders would announce a date for when the communities living nearby could harvest the fish from these pools. Harvesting involved all capable villagers, and not only professional fishers. Such harvests were not sold but used as food. On the bases of generosity and good neighbourliness, if the catch exceeded what a household could eat, they would send the remainder as gifts to relatives and people in other villages away from the pools. Some was dried or smoked and preserved for future use. Often, some fish were left in the pools as seed for the following year's harvest. Since most of these water bodies do not now exist, due to drought or changes to river topography, these practices have died out.

The decline of *konokono* stocks

Jarreng (about 237 km from the coast) and surrounding villages in the CRD are among the most active fishing bases in the division. Many of Jarreng's 55 fishers fish on a full-time basis. In the 1970s, fish were abundant and fishing was restricted in the dry season. During the rainy season, fishers engaged in farming. There were only a few migratory fishers who also went back home to farm when the rains approached. Late in the 1970s, however, there was a severe multi-year drought and many local people began turning to full-time fishing as an alternative livelihood. More migrant fishers also arrived. The population increased, markets expanded and, in response, fishing intensified with increased use of indiscriminate gear. Fishers rightly believe that year-round fishing has disrupted the breeding cycle of the *konokono*, and that stocks, especially of larger fish, have diminished. Some fishers began calling for closed seasons in order to protect breeding *konokono* populations.

For several years, fishers tried to create and implement fishing rules with little success. They sought assistance from local authorities and the heads of fishing communities, and formed small associations both within and between villages to tackle problems. But the rules that they set at the local level did not have legal backing and, even though they had been agreed upon, were not always respected. Because of this, conflicts occurred. Many fisher's associations simply collapsed.

The presence of migrant fishers only served to intensify these problems. Because indigenous fishers are also part-time farmers, they perceived migrant fishers as taking advantage of them, and fishing during the farming season when competition from locals was low. The 'strangers' settled in hidden bush camps ('*dakaa*') close to river tributaries that served as fish breeding grounds and contained mainly juveniles. These migrant fishers used destructive fishing gear, such as '*dumbou*' (woven baskets), small mesh nets and small hooks that caught fish indiscriminately and killed juveniles. In the meantime, many local fishers were obeying the rules. Under such circumstances, local fishers stopped obeying the rules, arguing that they could not watch while foreigners reaped the benefits of what they had patiently waited to mature.

Indigenous fishers often complain that the migrant fishers are irresponsible and selfish. 'Their interest is limited to catching the fish and going away, uncaring about when and where they fish, or believing that fish can never be finished,' explained one fisher, Jibi Sey. A foreign fisher himself, who settled and has fished in the area for more than three decades, Jibi Sey is deeply involved in community organisation and management. As a strong advocate of community participation and the responsible management of local fisheries resources he said: 'We depend on the fish resources for our livelihoods and therefore we ought to ensure that this fish is there for us today and for our children tomorrow, and as you know, that tomorrow only comes after we are dead.'

In an effort to assist with the resolution of these conflicts, the Fisheries Department and local authorities have occasionally intervened. Conflicts have occurred to which police, fisheries and district authorities have been called to mediate. One attempt to solve this problem was to ask all foreign fishers to abandon their bush camps and join their counterparts at the main fishing bases in the villages. However, most of them repeatedly returned to the bush camps, either deliberately or because they did not know about the rules.

In summary, local fishers at various locations in the CRD have tried, with little success, to control the exploitation of *konokono* resources. Their failure has often been due to the lack of understanding and awareness of the rules by foreigners and a lack of commitment, facilities, support and enforcement capabilities.

The CRDFA fisheries co-management strategy

Over the years, many fishers in the CRD have become conscious of their dependence on local fish resources and concerned about their continued availability. The fishers want to control fishing techniques that capture or injure parent and young *konokono*. Fishers agree that access to these breeding stocks must be restricted by some combination of direct effort control measures, such as closed seasons, gear restrictions and closed areas. Following repeated calls and efforts to establish closed fishing grounds/seasons, the fishing communities of Jarreng and other CRD fishing villages eventually formed the CRD Fishermen Association (CRDFA) in 1993/94. The formation of the Association was motivated by the realisation of the weaknesses involved when working as individuals, the desire to co-operate over the protection and conservation of *konokono* stocks in the area and to solve other common problems such as duplicated tax payments, fish marketing problems and the lack of technical and financial assistance, including credit facilities. The

Association comprises the fishers and fish traders of over 22 villages in the CRD area, including resident foreigners.

Each CRDFA member is expected to pay a membership fee of GD 20 (US$ 1.31) and is issued with a membership card. Members also make monthly cash contributions of GD 5 (US$ 0.33), which are recorded on their cards. Monies paid and contributed are collected by the cashier or treasurer and verified by the president of the Association before eventually being deposited into a bank account.

Virtually all professional fishers in the division are involved in CRDFA meetings. Through these, many fishers have learnt about fisheries management and the specific rules governing the *konokono* fishery. The CRDFA sought the assistance of the Fisheries Department (which had gained a lot of experience in working constructively with fisher and community groups through the implementation of the CFCs) and the Office of the Commissioner (the highest government authority in the area). Fisheries officials and the commissioner attend regular CRDFA meetings or, if not present, are later informed of the outcomes of the meetings. Depending on the nature of the actions required, Fisheries personnel either attend to them at the local level or communicate them to the Department. If necessary, local authorities from other sectors (e.g., health, agriculture and forestry) are approached for assistance to deal with fishers' problems/requests. All foreign fishers are expected to register with the Association and to comply with the rules, just like indigenous fishers. For locals, however, it is not obligatory to register with the Association.

The CRDFA convenes a special meeting at the beginning of every rainy season, to which all fishers in the area are invited. Here, they discuss ways and means to implement the closed season ('*hardi*'). The closures usually last from five to six months, from the beginning of the rainy season in June/July to late November/early December. Rules and fines are also decided upon at these *hardi* meetings, and then announced for all to hear. Other issues discussed include enforcement and policing rules. Previously, it was only one fisher that volunteered to police the Association's rules. He patrolled river tributaries in his fishing canoe (one of the few motorised canoes in the area) to observe and inspect fishing gear. Later, a number of selected fishing communities each identified two fishers responsible for policing and the enforcement of rules in their areas. In addition, every fisher is also supposed to enforce the rules during fishing operations.

The rules they are expected to enforce were agreed upon during *hardi* meetings:

(a) During the *hardi*, no fishing with hook and line may occur in the CRD.
(b) No net or hook and line fishing may occur in river tributaries during the *hardi*.

(c) Anybody caught breaking the rules will pay an agreed upon fine. If s/he fails to pay this, s/he will be taken before the chief's court (district tribunal) for settlement of the fine.
(d) All fishers must be based in the main fishing villages or camps and not in isolated bush camps.

The fine imposed on rule breakers in 1999 was set at GD 250 (US$ 16.34). Since the CRDFA's formation, 12 people have been caught and fined. Fines contribute towards the Association's funds and are used to cover expenses such as travel, meetings, organisation and logistics, including fuel for patrols. Initially, three fishers were caught and fined for fishing in river tributaries during the closed season. They refused to pay their fines, which provoked a lot of concern amongst members. Tension was so considerable at the time that the matter was referred to the District Chief's Court, where the offenders were each fined GD 500 (US$ 32.68) and threatened with imprisonment if they did not pay. The offenders did pay, although the money apparently went into government coffers. Fishers have, however, since developed respect for the rules and those caught have either paid up or have agreed to pay in the near future. Some fishers advocate the creation of an additional rule to enable the confiscation of offenders' fishing gear until they pay. Now, banned gear, such as *dumbou*, is destroyed.

Two to three weeks prior to the end of the closed season, the CRDFA convenes another special meeting at which a date is agreed upon for the opening of the *konokono* season (*'chokti'*). All fishers will have prepared for the anticipated harvest, and gear will have been improved, by, for example, increasing the number of hooks on a line. On the day of the *chokti* and for the following few days, fishing is intense and full of excitement as folk become involved in the production, processing, marketing and distribution of the fish.

There are great economic gains to be obtained from the conservation and sustainable exploitation of *konokono* and for the first time, a species-specific community-based fisheries management system has been instituted in The Gambia with a degree of success. Since this management system started, *konokono* catches are reported to be up to 50 - 150 kg per boat day, declining gradually to about 80 kg towards the end of *chokti*.

Problems and challenges

Efforts at species-specific (e.g., *konokono*) community-based fisheries resource management in The Gambia are limited to the river where the size of resource base and potential yield is virtually unknown for most species. The lack of good quality data on the resource makes sound management, enforcement and extension very difficult. Fishers are of the opinion that the present management system for *konokono* should be extended to other species, but biological information, levels of exploitation, knowledge of social factors and other related information are required before appropriate conservation measures can be developed. In addition, changes to the Gambia River's ecosystem arising from salt intrusion into its swamps, has narrowed the ecological boundaries and reduced the number of breeding grounds available to freshwater fish. Hence, it makes sense to link the management of fisheries resources with that of related resources and the environment.

Some members of CRDFA are of the opinion that they alone cannot make the rules and adequately implement them. Indeed, some fishers are pessimistic and anticipate a breakdown in the *konokono* management system. They feel that it is important to have a formal and legal framework through which they can operate. Fishers feel that the Fisheries Department should collaborate with them to agree on rules that have legal backing and institute a legal framework to make the system more workable. At present, the CRDFA relies, at the highest level, on the District Chief's Court to ensure that fines are paid. There are, however, no legal provisions on which the Chief – let alone the Association - can rely on to prosecute rule breakers. This makes it difficult to ensure compliance with informal regulations. Hence, Association members advocate for the incorporation of their rules into the law, and the presence of law enforcement authorities, like the Fisheries Department, at their meetings to advise them on relevant sections of the law and to provide the necessary advice to guide users of the resource.

Co-management tends to work better when the resource base is solid and does not require a reduction in fishing effort (Hara, 1999). It remains to be seen if the CRDFA system can work if the preservation of the resource base requires a future reduction in fishing effort. The rate of population growth in The Gambia is high and there is increasing entry into the fisheries sector. These factors are bound to affect the long-term sustainability of the resource. Despite the rules, the use of destructive gear continues, especially by dishonest fishers when they know there is no one to see them, and the CRDFA lacks sufficient means (e.g., patrol vessels and accessories), finance and legal backing to adequately enforce the regulations they pass. The management system will have to provide convincingly good incentives and economic benefits, such as rewards to fishers for their engagement and patience, if it is to be sustained. In addition, without the legal provisions or the knowledge of existing fisheries laws, neither the CRDFA nor the Chiefs' Courts can effectively enforce rules and fines. They therefore need to be empowered with legal provisions to undertake these new responsibilities.

Conclusions: perspectives

Inland fisheries resources in The Gambia have the potential to provide opportunities for income generation and improved nutrition. More and more, local fishing communities depend on inland water resources as a source of livelihood. It is important to assure the sustainable exploitation of these resources through both regulation and monitoring. It is, however, difficult for government agents to adequately monitor fisheries resources and enforce regulations in remote areas in the absence of adequate manpower and facilities. Fishing communities find it difficult to manage resources in the absence of the legal right to do so.

Resource management is difficult without the collaboration of its users who must also be aware of the implications of over-exploitation and the need for resource conservation. Under conditions of increased demographic and fishing pressure on the resource due to a wide array of social, political and economic factors, and in the absence of adequate management measures, the sustainable exploitation of fish resources is in question. It would therefore seem that a collaborative approach to management with clearly defined and agreed upon rules and regulations that can be effectively enforced by both government and communities (co-management), is one solution to the conservation of the fisheries resources. In many areas of The Gambia, including along most parts of the river, community participation in resource management is in one way or another a feature. However, more awareness of management systems, along with better organisation and institutionalisation are necessary for fishing communities to play a more meaningful role in management.

The government should work closely with the fishing communities to achieve these objectives. Responsive and institutionalised bodies need to be created which are representative of both the community and the government and which could be the basis for a participatory management system. It seems that fishers are the rightful users of inland fish resources and it is in their own interests that both they and the government seek to effect management measures to conserve these. Since *konokono* fishers have had the initiative to informally regulate the exploitation of the resource in a positive manner, it is only fit that the government lends support to their aspirations. In close collaboration with fishers, appropriate rules and regulations and detailed procedures can be identified and established through organisational and institutional frameworks. There are plans by the Fisheries Department to work with the CRDFA to employ participatory processes to identify key issues and solve constraints through small project and action planning initiatives to support the management initiatives of the CRDFA.

Linkages could also be made with regional and international projects and bodies for support of the initiatives. The regional Sustainable Fisheries Livelihoods Programme (SFLP), which covers 25 West African countries, sponsored by the UK Government's Department for International Development (DFID) and executed by the Food and Agriculture Organization of United Nations (FAO), is a potential partner for the improvement of the management system of the river fisheries (FAO, 1999c). The Programme centres on poverty alleviation through the improvement of the livelihoods of poor fishing communities. It uses the Sustainable Livelihoods Approach (SLA) and the FAO Code of Conduct for Responsible Fisheries (Barg, 2000) as tools and applies participatory approaches to the diagnosis of priority problems, identification of projects and implementation. The SFLP encourages community participation in fisheries resources management, along with structures and processes that improve livelihoods. Approaches are presently being followed towards the participatory identification of a small project for fishing communities along the Gambia River and the establishment of such structures and processes may provide support to local management initiatives. However, the government also has a big role to play in working with communities to create and support river fisheries management systems, which demand the government's commitment and support. Drawing on the experiences of the CFC concept and from the *konokono* management system should improve chances for success.

Acknowledgements

The authors are grateful to Messrs. Ahmed Khan and Randall E. Brummett for their improvement of earlier drafts of this paper. We also want to thank the Editors and unknown referees for their constructive criticism.

13

A Challenge Met? Some Final Thoughts

Kim Geheb and Marie-Therese Sarch

The papers contained in this volume leave little doubt that state-based management strategies on Africa's inland waters have failed. The very failure of the state in the managerial equation has created an adequate *raison d'être* for the development of alternative management strategies. The papers in this volume have examined various types of co-management, with varying degrees of community and government input, and considered what it is within their make-up that ensures that they might fail or succeed. At the same time, this volume has considered the kinds of trends and influences that might affect the outcome of these managerial regimes. Chief amongst these trends are both internal and external politics.

In this concluding chapter, we summarise the findings of this volume and pose answers to the questions originally posed in the introduction.

A synthesis of themes

Fisheries decline

There can be little doubt that the inland fisheries of Africa are on the decline. With the possible exception of the papers on Lake Kariba and The River Gambia, all of the articles in this volume present evidence to suggest that the fisheries they discuss are under considerable fishing pressure. At the same time, there appears to be little ameliorating influence from government sources. In view of the very grave problems faced by Africa's inland fisheries, the failure (for whatever reason) of the continent's fisheries departments to deliver even rudimentary conservation outputs must be considered tantamount to the 'do nothing' management option described by Reynolds *et al* in Chapter 7.

Trans-boundary problems

Having international boundaries traversing a fishery significantly impedes its management. The three African Great Lakes discussed in this volume are all shared resources, magnifying management problems and debilitating any attempt at co-ordinated management initiatives. These difficulties are only augmented when one of the lacustrine states is undergoing political turmoil, as is the case with the DRC on Lake Tanganyika, and the political instability around the north of Lake Chad. At far smaller levels, the success of community-based managerial initiatives on the shared lakes of Chilwa and Chiuta are undermined when shirkers can continue fishing illegally across the border from Malawi, in Mozambique.

The problem of size

In Chapter 2 Njaya suggests that one of the reasons for co-management working on Lakes Chiuta and Chilwa is that their ecosystems are small. Wilson's respondents (Chapter 9) also worried that it would be very difficult for a co-ordinated management strategy to occur across a lake as large as Lake Victoria. Generally, the common property resources literature argues that the smaller the resource and the fewer the number of exploiters, the better the chance that community-based solutions to its management may be found. Worrying about this may not, however, be an option for those concerned with Africa's heavily fished fish stocks. Provided management structures adequately nest local level institutions within broader co-ordinatory structures, then the problem may to some extent be minimised. In addition, if the unit around which management occurs is the landing site, the community or access[54], then the size of the fishery becomes largely irrelevant.

The access problem

Most of the fisheries described in this volume are placed within the context of serious socio-economic problems. It is possibly unsurprising, therefore, that fisheries should attract economic migrants be it because of depression in other sectors of the economy or because of the wealth that they promise. How access to fisheries manifests itself is, however, variable. In Chapter 9, Wilson describes fishes very much against any suggestion that access to the fishery be limited. At the same time, and concerning the same fishery, Medard *et al* (Chapter 10) consider the bigotry faced by a community of migrant fishers as they and established residents wrestle for control over a newly created managerial institution. The extent to which access is tolerated, therefore, appears to depend on circumstance. The discussion contained in N'Jie and Mikkola's paper (Chapter 12), Malasha's, Hara *et al.*'s (Chapter 3) and Njaya's (Chapter 2) all make reference to the very serious problems occasioned by the arrival of migrants at times when access to the resource, or the resource itself, is being viewed as threatened by lakeside communities.

---

[54] Notwithstanding the problems of defining these administrative units – see Agarwal and Gibson, 1999.

As Africa's population grows and becomes ever more youthful and at the same time, decades of 'development' appear not to have delivered the promise of wide-spread industrialisation, the employment of fishing labour in other sectors of the economy is no longer an option available to Africa's fisheries managers. Instead, the 'welfare function' of Africa's fisheries continues to increase, as they provide the un- and under-employed with some nutritional and income respite. This latter responsibility is daunting, for it must be weighed against the conservation losses that it necessarily implies. Reynolds *et al.*'s paper (Chapter 7) tries to address the worrying possibility that management successes on Lake Tanganyika might actually serve to attract more entrants, while Njaya and Hara *et al.* (Chapter 2 and 3) both suggest that a possible reason for the success of the Chiuta/Chilwa co-management programmes may be linked to the remoteness of these fisheries and the very lack of 'commercialisation' in the areas surrounding them. This latter perspective echoes Daly's (1991, 1992; Daly and Cobb, 1989) argument that growth economies contradict the objectives of conservation strategies, and questions the WCED's logic that poverty alleviation is positively correlated with satisfactory resource management and exploitation (World Commission on Environment and Development, 1987).

If communities of resource users are to assume or retain responsibilities for controlling access to fisheries, then their perceptions of the resource need to be understood. In several papers in this volume, communities have argued against migrant fishers assuming responsibilities for the fishery; migrants, they argue, cannot be expected to use the resource sustainably because they do not have a long term interest in the resource. Between communities and their resources, therefore, there exists an undefined moral responsibility, engendered by a history of association. How far this morality will actually extend, and whether or not it can form the basis of managerial action, is debatable.

This is not, however, a concern shared by governments, it seems. Hara *et al.* (Chapter 3) describe how '...it was assumed by the Fisheries Department that by making the user-community responsible for the management of their own fishery, then the problem of over-exploitation will be rectified by making them morally responsible for their negative exploitation patterns and illegal fishing activities. The guilt they feel could reduce illegal activities and make them more responsible towards the regulations'. This is equally true on Lake Victoria where governments believe that the enactment of law in parliament will somehow yield wholesale obedience at the lake shore, without any intervening act of enforcement.

National and international policies

The patchwork of donor and national policies and the resulting managerial befuddlement described by Allison and his colleagues (Chapter 4) makes for depressing reading. Their description of Malawi's new fisheries act suggests legislation designed to sufficiently appeal to donor sensibilities while adequately conservative to maintain the *status quo* of national fisheries administration. Any allusions to fisheries conservation may become entirely lost as (government or donor) actors pursue their respective political agendas. Nowhere is this made clearer than in Nyikahodzoi's paper, (Chapter 5) where national policies make not even a pretence of worrying about the messy details of fisheries conservation. Instead, Zimbabwe's Lake Kariba fishery is a stage on which government policies are being articulated at the expense of fisheries management. The role of international donors in this farce has not helped, and has ensured that 'Redressing the inequitable distribution of access rights to natural resources in Zimbabwe has degenerated into a conflict between the state and the western super powers' (Nyikahodzoi, this volume). A similar casualty is alluded to by Reynolds and his colleagues (Chapter 7), who remark on the apparent side-lining of the FAO-FINNIDA work on Lake Tanganyika by a new World Bank initiative with its equally new set of aspirations.

In some respects, decentralisation policies in Uganda have yielded similar outcomes. Here, Geheb and his colleagues (Chapter 8) describe how district fisheries officers now assume greater revenue collection responsibilities, with little or no accompanying law enforcement. This is also the case in the Tanzanian sector of Lake Victoria.

A related problem is how donors and governments define 'co-management', and how they think that it should be inspired. In Zambia's Lake Kariba fishery, a collaborative state-donor intervention assumed that co-management implied the creation of water-based user rights for lake-side communities, while in Malawi, the co-management initiative on Lake Malombe and the Upper Shire River may well have backfired because it was too dependent on donor-provided resources and inspiration.

Local-level political struggles

The importance of local-level political struggles is made patently clear in Krings' and Sarch's examination of the Nigerian Lake Chad fishery (Chapter 11). The interests they describe were sufficiently powerful to completely override federal fisheries management initiatives. Local-level manoeuvring and political struggles will also ensure that state regulation becomes interpreted and re-interpreted in terms of individual and community perceptions of their strategic position vis-à-vis the resource base. Wilson's analysis (Chapter 9) of how Tanzanian fishing communities interpret state regulations suggests that the way in which Dar Es Salaam-based law makers interpreted their own legislation may be wholly re-interpreted on Lake Victoria's shores.

What is clear from this discussion is that political struggles at the fisheries level may be as vicious as national-level political manoeuvring, particularly when new managerial initiatives involve the creation of novel institutions that may influence access to the fishery. Such institutions will, therefore, be strategically assessed, and actors will try to use these to improve or consolidate their resource access. The Kabangaja experience described by Medard and her colleagues (Chapter 10) relates a bitter conflict prompted by the introduction of Beach Management Units. Hara *et al.* (Chapater 3) place local political struggles squarely at the centre of their discussion on the success of Beach Village Committees (BVCs) in the management of a number of Malawi's fisheries. The development of these new institutions on Lakes Chiuta and Chilwa to some extent enabled resident communities to obtain sufficient power to eject migrant fishermen, while on Zambia's Lake Kariba fishery, a similar intervention enabled local Tongan fishing communities to impose their rule over immigrants.

How actors attempt to influence the outcome of such political struggles depends, Nyikahodzoi argues '...to a large degree on...organisational abilities...connections to the central political authority and...on the institutional order governing the distribution'. (Chapter 5) What should be noted is that these struggles may wholly fail to deliver any kind of conservation output, a point clearly made by Krings and Sarch (Chapter 11).

The introduction of new community-level institutions will almost certainly favour some groups above others, as both Nyikahodzoi and Wilson (Chapter 5 and 9) point out. In designing new management systems for these fisheries, therefore, it must be assumed that such struggles will affect the outcome of any intervention, and the objective then becomes one of trying to ensure that any resulting social or economic disequilibria may be minimised.

Government-community collaboration

The contributors to this volume by and large concede a role for the state in the management of Africa's inland fisheries. What role the state should play, however, is a contested topic. In Hara *et al.*'s paper, the *absence* of the state in the development and implementation of the co-management regime on Lakes Chiuta and Chilwa may be a reason for its success. Phrased differently, management regimes inspired, designed and implemented at the local level are far more likely to succeed than external interventions. The success of the CRD Fishermen's Association on the River Gambia may also, in large measure, be attributed to its origins amongst the river's resource users.

In part, many of these papers imply, the reasons why this should be the case is because such locally inspired regimes can better cope with the equally localised vagaries of social, economic and political life. Many of the papers contained within this volume suggest that external solutions aimed at solving very local problems (of whatever variety) are unlikely to work because of their origins in contexts worlds apart from small, riparian fishing communities. As Medard *et al.* comment on the Kabangaja conflict, "Outside solutions to conflicts so contingent on fluctuating internal socio-political dynamics are unlikely to work".

This is not, however, to suggest that governments should be excluded from the fisheries management process. N'Jie and Mikkola's examination of community-based management on the River Gambia suggests that communities are only willing to exercise their managerial powers up to a certain threshold. Geheb *et al.* propose that this threshold will be determined, on the one hand, by what the state allows community management institutions to do; and on the other hand, by what communities themselves will allow these organisations to do. In the latter case, community-based management organisations will not carry out managerial functions that lead them to being ostracised by their own communities. Geheb *et al.* report on work that suggests that communities' evaluation of offences will turn on who was involved, the circumstances under which the offence occurred and the gravity of the offence itself. The greater the severity of the punishment deemed necessary, the more likely it is that the community will seek outsiders to sanction offenders. It is at this junction that we perceive important roles for the state: to deal with those management questions which communities themselves feel they cannot.

Geheb and his colleagues argue (Chapter 8) that, for Lake Victoria, government involvement in the fishery's management should increase with the area to be managed. Hence, at regional levels, it makes sense to have the state involved to provide a co-ordinating role and to deal with matters between countries, lake basins etc. Reynolds *et al.*, (Chapter 7) whose work on Lake Tanganyika graphically portrays the need for governmental intervention in the management of this truly multi-national lake, also tackle this question of scale.

Geheb *et al.* go on to argue that because of the depressing history of practical regulatory delivery by the state, it is logical that the state should play a far greater role in extension in any future managerial intervention. They could provide advice on the state of the stock, limnological changes, new government decrees and fish market fluctuations, information that local-level managers will need if they are to fulfil their roles, and which will better equip them for any managerial negotiations that follow.

It is necessary that each stakeholder in a management plan exists to monitor and counteract the worst excesses of the other stakeholders. In this respect, therefore, government is also needed within a co-management plan if we are to avoid the kinds of excesses described by Krings and Sarch (Chapter 11).

The role of negotiation

The '*hardi*' meetings described by N'Jie and Mikkola (Chapter 13) involve all the fishermen of Jareng on the River Gambia. Government officials are also invited. At these meetings open and closed seasons are set, offences discussed and fines imposed. The size of the meetings suggests that these are open forums for negotiations to occur between community members and between the community and the state.

The dynamics of BVC activity in Malawi suggested daily negotiated outcomes between the BVCs and traditional authorities. The bribery that is so much a feature of government activities on Lake Victoria suggests that even government regulations are persistently subjected to negotiated outcomes. Malasha's portrayal of the Zambian Lake Kariba fishery (Chapter 6) suggests that those managerial outputs that the co-management regime did deliver were all derived from often intense negotiations between the actors and the various institutions involved. In short, negotiation as a means to arriving at decisions is a daily and integral feature of every one of Africa's inland fisheries.

It is for this reason that Geheb and his colleagues (Chapter 8) suggest a management structure based on forums within which negotiation can occur. Because impartial government regulation cannot be expected, then it makes sense to peruse some characteristic contained within the existing management regime, and exploit it as a management tool. The transaction costs of protracted negotiations are high, particularly if the output has to be based on consensus (Cleaver, 2000). These costs, however, may well be worth the output: universal consensus and management decisions that may not precisely meet the desires of all those involved, but at least partially do so.

Forums for negotiation, if formalised, may fall prey to Hara *et al.*'s observation (Chapter 3) that authoritarian traditional structures may well undermine democratic negotiation: 'Such historical and cultural contexts have relevance in that in a political environment in which people are not used to arguing or challenging government officials or their village headmen, giving people a formal platform for expressing their views and asserting their rights might not necessarily empower them to do so.' They are, however, a point at which to start, just as the 'socialisation' of Tanzania's BMUs is a start around which future co-managerial structures may be built (Medard and Geheb, 2000).

The legalisation of local-level input
As they attempt to end on an up-beat note, Allison *et al.* (Chapter 4) point out correctly that changes to the 1997 Malawian fisheries act legalising the role of fishing communities in the management of their resources is a laudable first step in the right direction. Geheb *et al.* (Chapter 8) praise the Ugandan Local Governments Act as a powerful tool that communities can use to augment their fisheries management roles. N'Jie and Mikkola (Chapter 12) stress the need for Gambian community fisheries management initiatives to be recognised formally in the laws of the state.

Many of the papers contained within this volume, therefore, consider the formal recognition of community-based management in legal doctrine to be *a priori* necessary for them to function adequately. This process must, however, proceed with caution, and every effort be made to avoid the debilitating ambiguities that Malasha identifies in the legal and administrative arrangements surrounding Zambia's management of the Lake Kariba *kapenta* fisheries. (Chapter 6)

The role of pre-existing institutions
In some cases, pre-existing institutions may undermine the efficiency of newly created ones, a point made by Hara *et al.* (Chapter 3) At the same time, Wilson (Chapter 9) argues that if any co-managerial system is to be implemented on Lake Victoria, it should really build on pre-existing institutions. The same point is made by Geheb *et al.*, who argue that beach committees, because they already exist, need to be an integral component of any management plan for this lake. To some extent, basing management structures on such institutions may ameliorate the homogeneity that these structures may assume. The creation of completely new institutions that fail to build on pre-existing local institutions, as has occurred on Lake Malombe and the Upper Shire River, and the Zambian Lake Kariba fishery, would appear to have created more problems than they have solved.

Answering questions

It will be recalled that in the introduction to this volume, we posed a number of questions concerning the management of Africa's inland fisheries. As we conclude this volume, we consider whether or not these questions can to any degree be answered given the trends identified above.

What are the reasons for the failures of state based management systems to achieve optimal yields from their inland fisheries?
Clearly, lack of staff, low levels of expertise, technological and funding problems, and insufficient legal and political backing all contribute to the

failures of fisheries departments across Africa to achieve the objectives of the management systems for which they are responsible. The frequent and possibly fundamental conflict between the objectives of centralised fisheries departments and those of fishing livelihoods is questioned throughout this volume. Adverse socio-political and economic environments throughout Africa ensure that entry into her inland fisheries has increased and their 'welfare function' maintained. The culmination of these experiences provides the background to most of the co-management initiatives described in this volume and underpins their proposals to redefine the roles of fisheries departments so as to emphasise support or extension activities.

Can Africa's customary and community-based management systems conserve fish stocks and sustain fishing livelihoods?

The objectives, whether implicit or explicit, of customary management systems rarely prioritise fisheries conservation, and the long-term sustainability of fishing livelihoods. Objectives more often focus on conflict prevention, profit, and survival – sustaining livelihoods in the long term is often overshadowed by more immediate needs for food and income. The prospects for the success of such systems in conserving fish stocks where they are already under intense pressure are poor. If customary systems are to conserve fish stocks and sustain fishing livelihoods, then fisheries managers need to understand community goals, and consider carefully whether or not it is possible to implement fisheries management in areas with low levels of food and personal security.

Important objectives for community-level institutions are mechanisms for resolving conflicts – this should be an important objective for all fisheries management systems, and those devising them need to identify where conflicts can arise and how they should be resolved. Where customary systems have resolved conflicts, this can often be attributed to powerful individuals rather than democratic systems. The question of leadership, therefore, looms large in the solution of conflicts and the identification of answers for fisheries management questions.

If the principal task of community-level institutions is to solve conflict, it will be necessary in the future to evaluate whether or not such institutions can cope with, or survive, having management responsibilities appended to their normal remits.

Should fisheries managers manage fish or humans?

Ultimately, the success of any management system depends on fishermen complying with management measures – given the widespread failure of fisheries managers to enforce these, the solution co-management offers is to persuade fishing communities to do this themselves – there are two broad alternatives for co-management strategies: either they ensure the convergence of objectives between different interest groups or the state provides incentives for fishers and/or power brokers to comply with their objectives. Either way, the future of co-management is dependent on an understanding of fishermen's objectives, their livelihood systems and the incentives they require. Biologists cannot provide this alone.

What does 'co-management' mean for Africa's inland fisheries?

The successful examples of co-management in this volume can be largely attributed to a convergence of fishing community and the state objectives. Successful co-operation seems to occur where each partner in this arrangement is able to draw upon the strength of the other in the pursuit of its objectives such as, for example, the exclusion of migrants from fisheries. Such convergence of objectives is, however, rare.

Community objectives are nearly always social ones (e.g. survival), and state objectives are nearly always conservation ones. The challenge for fisheries management is to identify some middle ground. Notwithstanding the complexities, the objectives of conservation are never likely to be completely be met and neither will community social objectives be entirely satisfied. A feasible goal is that in both instances, objectives will at least partially be achieved. A casualty of such arrangements may well be the precision hoped for in many former 'command-and-control' management systems. The 'fuzzy' output of a negotiated management regime may, however, compensate for this in that at least *some* management may occur, as opposed to the patent absence of management that characterises so many of Africa's fisheries today.

References

Aarnink, B. H. M. 1999. The politics of common resource management in Zambia's Mweru-Luapula fishery. In Hilhorst, T and Aaarnink, N. Co-managing the commons: setting the stage in Mali and Zambia. *Royal Tropical Institute Bulletin* 346. Amsterdam, Royal Tropical Institute: 41 – 62.

Adams, W. M. 1990. *Green development: environment and sustainability in the Third World.* London, Routledge.

Agarwal, A. and C. K. Gibson, 1999. Enchantment and disenchantment: the role of community in natural resource conservation. *World Development* 27 (4): 629 – 649.

Alexander, P. 1977. Sea tenure in Southern Sri Lanka, *Ethnology* 16 (1977): 231-251.

Allison, E. H., 2000. Sustainability in the African Great Lakes: It ain't what you do it's the way that you do it. In: Irvine, K., M. Munawar and R. Hecky (Eds.) *Sustainability of Great Lakes: Is it a scientifically sound and practical concept?* Proceedings of the GLOW II Symposium. Sligo, Ireland. July 18-21, 2000.

Allison, E. H., 2001. Big laws, small catches: global ocean governance and the fisheries crisis. *Journal of International Development* 13: (in press)

Allison, E. H., F. Ellis, P.M. Mvula and L. Mathieu, 2001a. Fisheries management and uncertainty: the causes and consequences of variability in inland fisheries in Africa, with special reference to Malawi. In: O. Weyl, (Ed.), *Proceedings of the National Fisheries Management Symposium, Lilongwe, Malawi, June 5-9th, 2001.* Published on CD-ROM, available from National Aquatic Resource Management Programme (NARMAP), Malawi.

Allison, E. H., K. Irvine, A. B. Thompson and B. P. Ngatunga, 1996. Diets and food consumption rates of the pelagic fish of Lake Malawi, Africa. *Freshwater Biology* 35: 489-515

Allison, E. H., R. G. T. Paley, G. Ntakimazi, V. J. Cowan and K. West, 2000. Biodiversity Assessment and Conservation in Lake Tanganyika: Final Technical Report to UNDP/GEF project, *Pollution Control and Other Measures to Protect Biodiversity in Lake Tanganyika* (RAF/92/G32). http://www.ltbp.org/BIOSS1.pdf

Allison, E., R. G. T. Paley, G. Ntakimazi, V. J. Cowan and K. West, 2001b. Final BIOSS technical report: Biodiversity and conservation in Lake Tanganyika. UNDP/GEF/RAF/92/G32. 183p.

Allison, E. H., G. Patterson, K. Irvine, A. B. Thompson and A. Menz, 1995. The pelagic ecosystem. In: A. Menz (Ed.), *The Fisheries potential and productivity of the pelagic zone of Lake Malawi/Niassa.* Chatham, UK, Natural Resources Institute: 351-377.

Allison, G. W., J. Lubchenco, and M. H. Carr, 1998. Marine reserves are necessary but not sufficient for marine conservation. *Ecol. Appl.* 8 (Suppl.): 79-82.

Allison, E. H., G. Patterson, K. Irvine, A. B. Thompson and A. Menz, 1995. The pelagic ecosystem. In: A. Menz (Ed.), *The Fisheries potential and productivity of the pelagic zone of Lake Malawi/Niassa.* Chatham, UK, Natural Resources Institute: 351-377.

Apostle, R. G., P. Barret, P. Holm, S. Jentoft, B. Mazany, B. McCay, and K. Mikalsen, 1998. *Community, state, and market on the North Atlantic rim: challenges to modernity in the fisheries.* Toronto, University of Toronto Press.

Bailey, C. and S. Jentoft, 1990. Hard choices in fisheries development. *Marine Policy* 14(4): 333-344.

Bailey, C., Cycon, D. and M. Morris, 1986. Fisheries development in the Third World: the role of international agencies. *World Development*, 14 (10/11): 1269-1275.

Balakrishnan, M and D. E. Ndhlovu, 1992. Wildlife utilization and local people: a case study in Upper Lupande game management area, Zambia. *Env. Conserv.* 19 (2): 135-143.

Baland, J.-M. and Platteau, J.-P. 1996. *Halting degradation of natural resources: is there a role for rural communities?* Oxford, Food and Agricultural Organization and the Clarendon Press.

Baland J. M. and J.P. Platteau 1999. The ambiguous impact of inequality on local resources management. *World Development* 27 (5): 451 - 482.

Banda, M. B. 1996. A report on the effectiveness of an increase in mesh size and closed season on kambuzi fishery in Lake Malombe. In Fisheries Department Progress Review of the Participatory Fisheries Management Programme for Lake Malombe and Upper Shire River (Proceedings of a workshop held at Boadzulu Lakeshore Resort, Mangochi, 27-29 August, 1996). Compiled by F. Njaya. Lilongwe, Malawi, Fisheries Department: 52-63.

Barg, U. 2000. The FAO Code of Conduct for Responsible Fisheries (CCRF). Paper presented in the First GEF Biennial International Waters Conference, Budapest, Hungary, October 14-18, 2000.

Barel, C. D. N., R. Dorit, P. H. Greenwood, G. Fryer, N. Hughes, P. B. N. Jackson, H. Kawanabe, R. H. Lowe-McConnell, M. Nagoshi, A. J. Ribbink, E. Trewavas, F. Witte and K. Yamaoka, 1985. Destruction of fisheries in Africa's Lakes. *Nature* 315, 19-20.

Barrow E. and M. Murphree, 1996 Community Conservation from Concept to Practice: A practical framework. In: Community Conservation in Africa: Principles and Comparative practices. *Institute for Development Policy and Management Working Paper* No. 8. Manchester, IDPM, University of Manchester.

Beadle, L.C. 1981. *The inland waters of tropical Africa* (2nd edition). London, Longman.

Bell, R. H. V. and S. J. Donda, 1993. *Community Participation Consultancy – Final Report, Volume 1: Community Fisheries Management Programme, Lake Malombe and Upper Shire River.* Lilongwe, Malawi, Department of Fisheries/MAGFAD.

Berkes, F. 1997. New and not-so-new directions in the use of the commons: co-management. *The Common Property Resource Digest.* No. 42. Indiana, International Association for the Study of Common Property.

Berry, S. 1989. Social institutions and access to resources. *Africa* 59 (1): 41 – 55.

Bertram, C. K. R, H. J. H. Borley and E. Trewavas, 1942. *Report on the Fish and Fisheries of Lake Nyasa.* London, Crown Agents for the Colonies.

Birley, M. H. 1982. Resource management in Sukumaland, Tanzania. *Africa* 52 (2): 1 - 29.

Blache, J. and F. Miton 1962. *Première contribution a la connaissance de la pêche dans le bassin hydrologique Logone-Chari-Lac Tchad.* Paris, ORSTOM.

Bland, S. J. R and S. J. Donda, 1994. Management Initiatives for the Fisheries of Malawi. *Fisheries Bulletin* No. 9. Lilongwe, Malawi, Fisheries Department.

Boe, T., 1998. Kapenta Production in the Sinazongwe area of Lake Kariba, Zambia. Paper presented to the Management, Co-management, No Management Workshop, Bergen, Norway, 10-12 September, 1998.

Boserup, E., 1965. *The conditions of agricultural growth: The economics of agrarian change under population pressure.* London, Allen and Unwin Ltd.

Bourdillion M. F. C. Cheater A. P. and Murphree M. W. 1985. *Studies of fishing on Lake Kariba.* Harare, Zimbabwe, Mambo Press.

Bundy A. and T. Pitcher 1995. An analysis of species changes in Lake Victoria: did the Nile perch act alone? In Pitcher, T. J. and P. J. B. Hart (Eds) *The impact of species changes in African lakes.* London: Chapman and Hall, Fish and Fisheries Series No. 18: 111 – 136.

Bwathondi, P. O. F. 1992. The fishery resource base for the Tanzanian sector of Lake Victoria. In *Report of a national seminar on the development and management of Tanzanian fisheries of Lake Victoria.* UNDP/FAO Regional Project for Inland Fisheries Planning (IFIP) RAF: 21-29.

Cacaud, P., 1996. Institutional choices for cooperation in fisheries management and conservation on Lake Tanganyika. Mission report. Fisheries Management and Law Advisory Programme, GCP/606/NOR/INT. Rome, FAO.

Cacaud, P., 1999a. Review of institutional and legal aspects relating to the management of Lake Tanganyika fisheries. Paper for presentation at the Eighth Session of the CIFA Sub-Committee for Lake Tanganyika. CIFA:DM/LT/99/Inf.6. Rome, FAO.

Cacaud, P., 1999b. Review of monitoring, control and surveillance system for Lake Tanganyika fisheries. Paper for presentation at the Eighth Session of the CIFA Sub-Committee for Lake Tanganyika. CIFA:DM/LT/99/Inf.7. Rome, FAO.

Campbell, J. and P. Townsley. 1996. Participatory and integrated policy: a framework for small-scale fisheries in Sub-Saharan Africa. Integrated Marine Management, United Kingdom.

Castro, A. P. 1997 Integrating Conflict Management into Forestry Policy: An applied Anthropologist's Perspectives. *Proceedings of a Satellite Meeting to the XI World Forestry Congress* 10-13 October, Anatolia, Turkey: Forest, Tree and People: Conflict Management Series: 195-208.

Chambers, R. 1983. *Rural development: putting the last first.* Harlow, Longman Scientific and Technical.

Chambers, R. and Conway, G. R. 1992. Sustainable rural livelihoods: practical concepts for the 21st Century. *I. D. S. Discussion Paper* 296. Brighton, Institute of Development Studies, University of Sussex.

Chapman, M. D. 1989. The political ecology of fisheries depletion in Amazonia. *Environmental Conservation* 16 (4): 331-337.

Charnley, S. 1997. Environmentally-displaced peoples and the cascade effect: lessons from Tanzania. *Human Ecology* 25 (4): 593 – 618.

Chilowa, W. 1998. The impact of agricultural liberalisation on food security in Malawi. *Food Policy* 23 (6): 553-569.

Chipungu P. and H. Moinuddin, 1994. Management of the Lake Kariba inshore fisheries (Zambia): a proposal. *Project Report* 32. Chilanga, Zambia/ Zimbabwe SADC Fisheries Project.

Chirwa, W. C., 1996. Fishing rights, ecology and conservation along southern Lake Malawi, 1920-1964. *African Affairs*, 95 (380): 351-377.

Chitembure, R. M., 1995. 1995 *Frame survey report.* Chilanga, Department of Fisheries.

Chitembure, R. M, L. Karenge, and M. Mugwagwa. 1997. 1996 Joint fisheries statistical report, Lake Kariba. *Project Report* No. 51. Kariba, Zambia/ Zimbabwe SADC Fisheries Project.

Chitembure, R. M., N. Songore and A. Moyo, 1999. 1998 Joint fisheries statistical report: Lake Kariba. *Project Report* No. 59. Kariba, Zambia/ Zimbabwe SADC Fisheries Project.

Cleaver, F. 2000. Moral ecological rationality, institutions and the management of common property resources. *Development and Change* 31 (2000): 361-383.

Coenen, E. J. (Ed.), 1994. Historical data report on the Fisheries, Fisheries Statistics, Fishing Gears and Water Quality of Lake Tanganyika (Tanzania). FAO/FINNIDA Research for the Management of the Fisheries on Lake Tanganyika. GCP/RAF/271/FIN-TD/15: 115p.

Coenen, E. J. (Ed.), 1995. LTR's fisheries statistics subcomponent: March 1995 update of results for Lake Tanganyika. FAO/FINNIDA Research for the Management of the Fisheries on Lake Tanganyika GCP/RAF/271/FIN/ TD/32 (En): 45 p.

Coenen, E. J. and E. Nikomeze, 1994. Results of the 1992-93 Catch Assessment Surveys. FAO/FINNIDA Research for the Management of the Fisheries on Lake Tanganyika. GCP/RAF/271/FIN-TD/24 (En) 22p.

Coenen, E. J., P. Paffen, and E. Nikomeze, 1998. 'Catch per unit of effort (CPUE) study for different areas and fishing gears of Lake Tanganyika. FAO/FINNIDA Research for the Management of the Fisheries of Lake Tanganyika. GCP/RAF/271/FIN - TD/80 (En).

COFAD, 2002. *Back to basics. Traditional inland fisheries management and enhancement systems in Sub-Saharan Africa and their potential for development.* Tutzing, Germany, Deutsche Gasellschaft fur Technisce Zusammerarbeit.

Cohen, A. S., 1991. Report on the First International Conference on the Conservation and Biodiversity of Lake Tanganyika. March, 1991. Bujumbura, Burundi, Biodiversity Support Program.

Cole, J. E., R. B. Dunbar, T. R. McClanahan and N. A. Muthiga, 2000. Tropical Pacific forcing of decadal SST variability in the Western Indian Ocean over the past two centuries. *Science* 287: 617-619.

Colson E.,1971. *The Social Consequences of Resettlement.* Manchester, Manchester University Press.

Conway, G. R. and Barbier, E. B. 1990. *After the green revolution: sustainable agriculture for development.* London, Earthscan Publications Ltd.

Copes, P. 1996. Adverse impacts of the individual quota system on conservation and fish harvest productivity. *Institute of Fisheries Analysis Discussion Paper* 96. Simon Fraser University, British Columbia, Canada

Corten, A, 1996. The widening gap between fisheries biology and fisheries management in the European Union. *Fisheries Research* 27 (1996): 1-15.

Coulter, G. W., 1970. Population changes within a group of fish species in Lake Tanganyika following their exploitation. *J. Fish Biol.* 2: 329-353.

Coulter, G. W., 1994. Lake Tanganyika. In: Martens, K., B. Goddeeris, and G. Coulter (Eds.) *Speciation in ancient Lakes*. Archiv fur Hydrobiolgie 44: 13 -18.

Coulter, G. W., 1999. Sustaining both biodiversity and fisheries in ancient lakes: the cases of Lakes Tanganyika, Malawi/Nyasa and Victoria. In: H. Kawanabe, G. W. Coulter, and A. C. Roosevelt (Eds.) *Ancient Lakes: Their cultural and biological diversity*. Belgium: Kenobi Productions: 177 – 187.

Coulter, G. W. (Ed.), 1991. *Lake Tanganyika and its life*. Oxford, British Museum (Natural History) and Oxford University Press.

Crean, K. and Geheb, K. 2001. Sustaining appearances: sustainable development and the fisheries of Lake Victoria. *Natural Resources Forum* 25 (3) 215 – 224.

Crean, K. and Symes, D. (Eds.) 1996. *Fisheries management in crisis*. Oxford, Fishing News Books, Oxford Science, Ltd.

Cycon, D. E, 1986. Managing fisheries in developing nations: a plea for appropriate development. *Natural Resources Journal* 26 (1): 1-14.

Daly, H. E. 1991. Sustainable growth: an impossibility theorem. *National Geographic Research and Exploration* 7 (3): 259-265.

Daly, H. E. 1992. *Steady-state economics*. London, Earthscan Books. Daly, H. E. and Cobb, J. B. Jr. 1989. *For the common good: redirecting the economy towards community, the environment and a sustainable future*. London, Green Print.

Davies, S. 1996. *Adaptable Livelihoods. Coping with Food Insecurity in the Malian Sahel.* London: Macmillan.

Department of Fisheries, 1995. Minutes of a meeting called by the Member of Parliament at Sinazongwe Fisheries Training Centre, Sinazongwe, 30/6/95. Sinazongwe, Department of Fisheries.

Derman, B. and Ferguson, A. 1995. Human rights, environment, and development: the dispossession of fishing communities on Lake Malawi. *Human Ecology* 23 (2): 125 – 142.

Desloges, C. 1997 Community Forestry and Forest Resource Conflicts: An Overview. *Proceedings of a Satellite Meeting to the XI World Forestry Congress* 10-13 October, Anatolia, Turkey: Forest, Tree and People: Conflict Management Series: 31-51.

Dissi, C. S and F. J. Njaya. 1995. the development of a community based fisheries management on Lake Chiuta. Unpublished paper presented at the Fisheries Co-management Workshop from October 28-30, 1995, Kariba, Zimbabwe.

Donda, S. J. 1996. The management of artisanal fisheries in Lake Malombe and Upper Shire. In: Njaya, F. J. (Ed.) The Progress Review of the

Participatory Fisheries Management Programme for Lake Malombe and Upper Shire. Unpublished document; Lilongwe, Malawi, Fisheries Department: 14-19.

Donda, S. J. 2001. Theoretical advancement and institutional analysis of fisheries co-management in Malawi: experiences from Lakes Malombe and Chiuta. Unpublished Ph.D. Thesis. Aalborg, University of Aalborg.

Dunn, I. G. and Ssentongo, G. W. 1992. *Regional framework for the management of the fisheries of Lake Victoria.* UNDP/FAO Regional Project for Inland Fisheries Planning (IFIP). RAF/87/099-TD/46/92 (En). Rome, Food and Agricultural Organization, Rome.

Dustin Becker, C. and E. Ostrom, 1995. Human ecology and resource sustainability: the importance of institutional diversity. *Annual Review of Ecology and Systematics* 26 (1995): 113-133.

Ellis, F. 2000. *Rural livelihoods and diversity in developing countries.* Oxford, Oxford University Press.

Ellis, F. 2001. Emerging issues in rural development. *Development Policy Review* (in press)

FAO, 1982. *Fisheries Expansion Project, Malawi: Biological Studies on the Pelagic Ecosystem of Lake Malawi.* FI:DP/MLW/75/019 Technical Report 1. Rome, Food and Agricultural Organization/United Nations Development Program.

FAO, 1993. Fisheries Management in south-east Lake Malawi, the Upper Shire River and Lake Malombe. *CIFA Fisheries Technical Report* 21. Rome, Food and Agricultural Organization.

FAO, 1994a. *Inland Fisheries Planning, Development and Management in Eastern-Central-Southern Africa. Terminal Report: Project Findings and Recommendations.* FI: DP/RAF/87/099. Rome, Food and Agricultural Organization/United Nations Development Program. 42p.

FAO, 1994b. Fisheries characteristics of the shared lakes of the East African rift. *CIFA Technical Paper* Vol. 24, Food and Agricultural Organization, Rome: 27-28.

FAO, 1995a. Management of African inland fisheries for sustainable production. Paper presented at the First Pan African Fisheries Congress and Exhibition, UNEP, Nairobi. Rome, Food and Agricultural Organization.

FAO, 1995b. *Code of conduct for responsible fisheries.* Rome, Food and Agricultural Organization.

FAO, 1995c. *Chambo Fisheries Research, Malawi: Project Findings and Recommendations.* FI:DP/MLW/86/013. Rome, Food and Agricultural Organization.

FAO, 1996a. *Technical guidelines for responsible fisheries, No. 2: Precautionary approach to capture fisheries and species introductions* Rome, Food and Agricultural Organization.

FAO, 1996b. Fisheries and Aquaculture in Sub-Saharan Africa: Situation and Outlook in 1996. *FAO Fisheries Circular* No. 922 FIPP/C922. Rome, Food and Agricultural Organization.

FAO, 1997. *Technical guidelines for responsible fisheries, No. 4: Fisheries management*. Rome, Food and Agricultural Organization.

FAO, 1999a. Fishery Country Profile. FID/CP/Gam Revision 3: 1-4. Rome, Italy, Food and Agricultural Organization.

FAO, 1999b. Fishery Statistics – capture production. FAO Yearbook Vol. 88/1:1-754. Rome, Italy, Food and Agricultural Organization.

FAO, 1999c. Sustainable Fisheries Livelihoods (SFL) Programme. FAO/Government Co-operative Programme. Project Document GCP/INT/735/UK. Rome, Italy, Food and Agricultural Organization: 1-39.

FAO, 2000. World Capture Production: Fish, Crustaceans, molluscs etc. Table A-1 (a). ftp://ftp.fao.org/fi/stat/summ–99/capt–a-/a.pdf

Federal Government of Nigeria, 1992. Inland Fisheries Decree *Supplement to the Official Gazette Extraordinary No 75, vol. 79*, 31st December 1992.

Ferguson, A. E. and B. Derman, 2000. Writing against hegemony: development encounters in Zimbabwe and Malawi. In: Peters, P. (Ed) *Development Encounters. Sites of Participation and Knowledge*. Harvard, Harvard Institute for International Development: 121-156.

Ferguson, A. E, B. Derman and R. M. Mkandawire, 1993. The new development rhetoric and Lake Malawi. *Africa* 63: 1-18.

Fisheries Department. 1993. *Artisanal Fisheries Management Plan*. Malawi, Lilongwe, Fisheries Department.

Fisheries Department, 1997. Annual Frame Survey (Unpublished). Lilongwe, Malawi, Fisheries Department.

Fisheries Department. 1997. *A Guide to the Fisheries Conservation and Management Act 1997*. Lilongwe, Malawi Fisheries Department.

Fogarty, M. J., 1999. Essential habitat, marine reserves, and fishery management. *Trend Ecol. Evol.* 14: 133-134.

Forman, S. 1967. Cognition and the catch: the location of fishing spots in a Brazilian coastal village. *Ethnology* 6 (1967): 417-426.

Fryer, G., 1999. Local knowledge of the fishers of the ancient lakes of Africa, and an example of comprehensive understanding. In: Kawanabe, H., G. W. Coulter and A. C. Roosevelt (Eds.) *Ancient Lakes: Their cultural and Biological Diversity*. Ghent, Belgium, Kenobi Productions: 363-270.

FSTC (Frame Survey Technical Committee), 2001. Draft report on Lake Victoria fisheries frame survey 2000. Part 1: Main Report. Jinja, Lake Victoria Fisheries Organization Secretariat.

Geheb, K. 1997. *The Regulators and the regulated: fisheries management, options and dynamics in Kenya's Lake Victoria Fishery*. Unpublished D.Phil. Thesis. Falmer, Brighton, University of Sussex.

Geheb, K., 1999. Small-scale regulatory institutions in Kenya's Lake Victoria fishery: Past and Present. In: Kawanabe, H., G. W. Coulter and A. C. Roosevelt (Eds), *Ancient Lakes: Their Cultural and Biological Diversity*. Ghent, Belgium, Kenobi Productions: 113-122.

Geheb, K. 2000a. Fisheries legislation on Lake Victoria: present legislation and new developments. In K. Geheb and K. Crean (Eds.) 2000. The Co-management Survey: Co-managerial perspectives for Lake Victoria's fisheries *LVFRP Technical Document* No. 11. LVFRP/TECH/00/11. Jinja, The Socio-economic Data Working Group of the Lake Victoria Fisheries Research Project: 172 - 183.

Geheb, K. (Ed.) 2000b. The Co-management Survey: PRA reports from five beaches on Lake Victoria. *LVFRP Technical Document* No. 9. LVFRP/TECH/00/09. The Socio-economic Data Working Group of the Lake Victoria Fisheries Research Project, Jinja.

Geheb, K. and Crean, K. (Eds) 2000. The Co-management Survey: Co-managerial perspectives for Lake Victoria's fisheries. *LVFRP Technical Document* No. 11. LVFRP/TECH/00/11. The Socio-economic Data Working Group of the Lake Victoria Fisheries Research Project, Jinja.

Geheb, K., Crean, K., Odongkara, O. K., Abila, R., Medard, M. Omwega, R., Yongo, E., Onyango, P., Mlahagwa, E., Atai, A., Gonga, J., Nyapendi, A. T., Kyangwa, M., Lwenya, C., Nyamwenge, C., Zenge, B. and Wabeya, U. 2000a. The co-management survey: results of the survey of fishers. In K. Geheb and K. Crean (Eds.) 2000. The Co-management Survey: Co-managerial perspectives for Lake Victoria's fisheries. *LVFRP Technical Document* No. 11. LVFRP/TECH/00/11. Jinja, The Socio-economic Data Working Group of the Lake Victoria Fisheries Research Project: 49-62.

Geheb, K., Crean, K., Odongkara, O. K., Abila, R., Medard, M. Onyango, P., Omwega, R., Yongo, E., Mlahagwa, E., Atai, A., Gonga, J., Nyapendi, A. T., Kyangwa, M., Lwenya, C., and Nyamwenge, C., 2000b. The co-management survey: results of the survey of Fisheries Department personnel. In Geheb, K. and Crean, K. (Eds.) 2000. The Co-management Survey: Co-managerial perspectives for Lake Victoria's fisheries. *LVFRP Technical Document* No. 11. LVFRP/TECH/00/11. Jinja, The Socio-economic Data Working Group of the Lake Victoria Fisheries Research Project: 63 - 64.

Geheb, K., Abila, R., Onyango, P, Mlahagwa, E, Nyapendi, A. T, Gonga, J., Atai, A., Bwana, E. and Onyango, J. 2000. Report of the PRA carried out at Ihale Beach, Tanzania, April 11-14, 2000c. In Geheb, K. (Ed.) The Co-management Survey: PRA reports from five beaches on Lake Victoria. *LVFRP Technical Document* No. 9. LVFRP/TECH/00/09. The Socio-economic Data Working Group of the Lake Victoria Fisheries Research Project, Jinja: 5-35

Gibbon, P. 1997. Of saviours and punks: the political economy of the Nile perch marketing chain in Tanzania. *CDR Working Paper* 97.3. Copenhagen, Centre for Development Research, June 1997.

Goldschmidt, T., Witte, F. and Wanink, J. 1993. Cascading effects of the introduced Nile perch on the detrivorous/phytoplantivorous species in sublittoral areas of Lake Victoria. *Conservation Biology* 7 (3): 686-700.

Government of Malawi. 1987. *Statement of Development Policies*. Malawi Government. Zomba.

Government of Malawi. 1997a. The Fisheries Conservation and Management Act. Zomba, Government Printer.

Government of Malawi, 1997b, *Malawi Decentralization Policy*. Zomba, Malawi: Government Printer.

Government of Malawi, 1997c. Fisheries Conservation and Management Regulations, Malawi Gazette Supplement dated 23rd June 2000 (published 14th July 2000),

Government of Malawi, 1999. *Fish Stocks and Fisheries of Malawian Waters – Resource Report 1999*. Lilongwe, Malawi, Fisheries Department, Government of Malawi.

Government of Malawi, 2000. *National Development Report, 1999*. Lilongwe, Malawi: Economic Policy Research Centre, Office of the President and Cabinet, Government of Malawi.

Government of Malawi/United Nations Development Programme. 1993. *Situation Analysis of Poverty in Malawi*. Lilongwe, Government of Malawi..

Government of Malawi/United Nations Development Programme. 1998. Management for Development Programme: Revised Mangochi socio-economic Profile. Mangochi, Malawi.

Government of the Republic of Kenya, Government of the Republic of Uganda and Government of the United Republic of Tanzania. 1995. *Lake Victoria Environment Management Program Proposal*. Submitted to the World Bank, November 1995.

Government of Zimbabwe. 1981. *Growth with equity: an economic policy statement*. Harare, Zimbabwe, Government Printers.

Gréboval, D. and Fryd, D. 1993. *Inland fisheries of Eastern/Central/Southern Africa: basic fisheries statistics*. FAO/UNDP (Bujumbura), June 1993, Ref.: RAF/87/099-TD/52/93 (En.). Rome, Food and Agricultural Organization.

Gréboval, D., M. Bellemans, and M. Fryd, 1994. Fisheries characteristics of the shared lakes of the East African Rift. *CIFA Technical Paper* 21. Rome, Food and Agricultural Organization.

Hachongela, P., J. Jackson, I. Malasha, and S. Sen (Ed.), 1998. Analysis of emerging co-management arrangements in the Zambian inshore fisheries of Lake Kariba. In Normann, A. K., J. R Nielsen and S. Sverdrup (Ed.s). *Fisheries Co-management in Africa.* Proceedings from a regional workshop on fisheries co-management, Mangochi, Malawi, 18-20 March, 1997. Hirtshals, Institute for Fisheries Management and Coastal Community Development, Hirtshals, 1998.

Hamid-Abdul, Y. 1997 The Role of Positive Conflict Management in Fostering Social and Political Changes: The Case of Enda Graf Forum and Renapop, Senegal. *Proceedings of a Satellite Meeting to the XI World Forestry Congress* 10-13 October, Anatolia, Turkey: Forest, Tree and People: Conflict Management Series: 149-165.

Hanek, G., 1994. Management of Lake Tanganyika fisheries. FAO/ FINNIDA Research for the Management of the Fisheries on Lake Tanganyika GCP/ RAF/271/FIN-TD/25 (En): 21 pp.

Hanek, G. and J. F. Craig (Eds.), 1996. Report of the Fifth Joint Meeting of the LTR's Coordination and International Scientific Committees. FAO/ FINNIDA Research for the Management of the Fisheries of Lake Tanganyika. GCP/RAF/271/FIN - TD/55 (En).

Hanna, S. 1995. Efficiencies of user participation in natural resources management. In Hanna, S. and M. Munasinghe (Eds.). *Property rights and the environment:: social and ecological issues*. Washington D. C.: 59-67.

Hanna, S. S., 1999. Strengthening governance of ocean fishery resources. *Ecological Economics* 31: 275- 286.

Hara, M. 1996. Problems of introducing community participation in fisheries management: lessons from Lake Malombe and Upper Shire River (Malawi) Participatory Fisheries Management Programme. *Southern African Perspectives* No. 59. Belville (RSA), Centre for African Studies, School of Government, University of Western Cape.

Hara, M., 1998. Problems of introducing community participation in fisheries management: Lessons from the Lake Malombe and Upper Shire River (Malawi) participatory fisheries management programme. In: Normann, A. K., J. Raakjær Nielsen and S. Sverdrup-Jensen, (Eds.), 1998. *Fisheries Co-management in Africa. Proceedings from a regional workshop on fisheries co-management research in Mangochi, Malâwi, 18-20 March 1997*. Fisheries Co-Management Research Project, Research Report 12. ICLARM/IFM: 41- 60.

Hara, M. 1999. Fisheries co-management: a review of the theoretical basis and assumptions. *Southern African Perspectives* No. 77. Belville (RSA), Centre for African Studies, School of Government, University of Western Cape.

Hara, M. 2001. *Could co-management provide a solution to the problems of artisanal fisheries management on the Southeast Arm of Lake Malawi?* Unpublished Ph.D. thesis, Cape Town, South Africa, University of the Western Cape.

Hara, M. and E. Jul Larsen. Forthcoming. Lords of Malombe: an analysis of the development of effort in Lake Malombe. Report for The Management, No-management, or Co-management? Project. Bergen, Christian Michelsen Institute.

Harrigan, J. 1988. Malawi: The impact of pricing policy on smallholder agriculture 1971-88. *Development Policy Review* 6: 515-433.

Harris, C. K. 1998. Social regime formation and community participation in fisheries management. In Pitcher, T. J., Hart, P. J. B. and Pauly, D. (Eds.). *Reinventing fisheries management.* London, Chapman and Hall: 261-276.

Harris, P. G. 1942. The Kebbi Fishermen (Sokoto Province, Nigeria). *Journal of the Royal Anthropological Society* 72 (1): 23-31.

Hart, T., C. Imboden, D. Ritchie and F. Swartzendruber, 1998. Biodiversity conservation projects in Africa: lessons learnt from the first generation.. *Environment Department Dissemination Notes No. 62*, World Bank, Washington D.C.

Hartmann, W., 1987. *Socio-economics sectoral paper. Malawi Fisheries Development Strategy Study.* Lilongwe, Malawi: Unpublished report to Department of Fisheries by GOPA Consultants Ltd.

Herald, 13 June 1997. Sliding Scale fees invalid: Court. Harare, Zimbabwe, *The Herald.*

Hersoug, B. 1996. Same procedure as last year, same procedures as every year: some reflections on South Africa's new fisheries policy. Unpublished paper.

Hersoug, B. and P. Holm 1998. Change without redistribution: an institutional perspective on South Africa's new fisheries policy. Unpublished paper.

Hersoug, B. and Holm, P. 1999. Bringing the state back in the choice of regulatory system in south Africa's new policy. Tromsø, Norway, University of Tromsø, Unpublished paper.

Hersoug, B and S.A. Rånes, 1997. What is good for the fishers is good for the nation: Co-management in the Norwegian fishing industry in the 1990s. *Ocean and Coastal Management*, 35 (2-3): 157-172.

Hilborn, R., Walters, C. J. and Ludwig, D. 1995. Sustainable exploitation of renewable resources. *Annual Review of Ecology and Systematics* 26 (1995): 45-67.

Hill, P. 1972. *Rural Hausa: a village and a setting.* Cambridge, Cambridge University Press.

Hoekstra, T. M., Asila, A., Rabour, C. and Rambiri, O. 1991. *Report of a census of fishing boats and gear in the Kenyan waters of Lake Victoria.* FAO/UNDP Regional Project for Inland Fisheries Planning Development in Eastern/Central/Southern Africa, Bujumbura (Burundi): FAO/UNDP RAF/87/099-TD/26/91 (En.). Rome, Food and Agricultural Organization.

Holden, M. J. 1961. Fishing methods in Sokoto Province, N. Nigeria. *The Nigerian Field* 26 (4): 146-158.

Holm, P. B. Hersoug and S. A. Rånes 2000. Revisiting Lofoten: co-managing fish stocks or fishing space *Human Organization* 59 (3): 353-364

Hoza, R. and Mahatane, A. T. 1998. *Co-management in Mwanza Gulf.* Lake Victoria Environment Management Project, Fisheries Management Component. Dar es Salaam, Ministry of Natural Resources and Tourism, July 1998.

Hutton, J. M. 1991. *The fisheries of Lake Kariba.* Zambia-Zimbabwe SADC Fisheries Project Consultants' Report Document 10. Harare, Zimbabwe

ICLARM/GTZ. 1991. The context of small-scale integrated agriculture-aquaculture systems in Africa: a case study of Malawi. Manila and Germany, ICLARM and the GTZ.

ICLARM/IFM. 1998. Analysis of co-management arrangements in fisheries and related coastal resources: a research framework. Unpublished MS. Manila, International Centre on Living Aquatic Resource Management.

Ita, E. O. 1993. *Transactions of the Inland Fisheries Laws and Regulations Drafting Committee, 25th April 1985.* New Bussa, Nigeria, National Institute for Freshwater Fisheries Research.

Jackson, P. B. N., T. D. Iles, D. Harding and G. Fryer, 1963. *Report on the Survey of Northern Lake Nyasa 1954-55.* Zomba (Nyasaland), Government Printer.

Jentoft, S. 1989. Fisheries co-management: delegating government responsibility to fishermen's organisations. *Marine Policy* 13 (April 1989): 137-154.

Jentoft, S. 1993. *Dangling Lines.* Newfoundland, Newfoundland Institute of Social and Economic Research.

Jentoft, S. and B. McCay. 1995. User participation in fisheries management: lessons drawn from international experiences. *Marine Policy* 19 (3): 227-230.

Jentoft, S., B. J. McCay and D. C. Wilson 1998. Social theory and fisheries co-management. *Marine Policy* 22 (4-5): 423-436.

Johannes, R. E. 1981. *Words of the lagoon: fishing and maritime lore in the*

Palau District of Micronesia. Berkeley, University of California Press.

Joint Zonal Management Committee, 1995. Minutes of the Joint Zonal Management Committee meeting to discuss the allocation of islands to artisanal fishermen held at Sinazongwe Fisheries Training Centre on 27/10/1995. Sinazongwe, Department of Fisheries: 1-22.

Jul-Larsen, E., 2000. Developments in Southern African freshwater fisheries: is co-management a solution? Draft paper presented at the International Symposium on Contested Resources: Challenges to Governance of Natural Resources in Southern Africa. University of the Western Cape, 18 – 20 October 2000.

Jumpha, D. 1996. The Impact of the fisheries co-management programme on the enforcement activities around Lake Malombe and Upper Shire. In: Fisheries Department (compiled by F. Njaya). Progress Review of the Participatory Fisheries Management Programme for Lake Malombe and Upper Shire River. Proceedings of a workshop held at Boadzulu Lakeshore Resort, Mangochi, 27-29 August, 1996). Lilongwe, Malawi, Fisheries Department: 33-36.

Kalk, M., A. J. McLachlan and C. Howard-Williams, 1979. Lake Chilwa: studies of change in a tropical ecosystem. *Monographae Biologicae*. 35 (1979): 17-227.

Kaluwa, B., E. Silumbu, E. Ngalande Banda and W. Chilowa, 1992. The structural adjustment programme in Malawi: A case of successful adjustment? In: Mwanza, A. M. (Ed) *Structural Adjustment in SADC: Experiences and Lessons from Malawi, Tanzania, Zambia and Zimbabwe*. Harare: SAPES Books: Chapter 2.

Kandoole, B. F., 1990. *Structural Adjustment in Malawi: Short Run Gains and Long Run Losses*. Zomba: University of Malawi, Chancellor College.

Kanyerere, G. Z., 1999. *Demersal exploratory fisheries research survey in Central and Northern Lake Malawi, 1998*. Bulletin of Fisheries No 41. Lilongwe, Malawi: Department of Fisheries.

Karenge, L. P., 1992. *Inshore fish population changes at Lakeside, Kariba between 1969-1991*. Unpublished M.Phil Dissertation, Bergen, University of Bergen.

Keohane, R. O. and E. Ostrom, (Eds.) 1995. *Local commons and global interdependence: heterogeneity and cooperation in two domains*. London, Sage Publications.

Kiiza, F. X. M. 1998. A presentation of the fisheries management issues at he research-stakeholder workshop on the fisheries resources of Lakes Kyoga and Victoria. In Proceedings of the stakeholders workshop on: the fisheries of Lake Victoria and Kyoga: Towards sustainable development and management of the fisheries resources of Lake Victoria and Kyoga. Jinja, 24th - 28th September, 1998. Jinja, Fisheries Research Institute: 54-59.

Kolawole, A. 1986. *Irrigation and Drought in Borno, Nigeria*. Ph.D. Thesis submitted to the University of Cambridge, 1986.

Kolawole, A. 1987a. Environmental change and the South Chad Irrigation Project (Nigeria). *Journal of Arid Environments* 13: 169-176.

Kolawole, A. 1987b. Cultivation of the floor of Lake Chad: a response to environmental hazard in Eastern Borno, Nigeria. *The Geographical Journal* 154 (2): 243-250.

Krings, M. 1998. Migrant Hausa communities in the lake: preliminary research notes from Lake Chad. *Borno Museum Society Newsletter* 34/35: 31-41.

Krings, M. 2000. Small fish, big money: Conflicts evolving around new fishing techniques and old fishing rights at the shores of Lake Chad / Nigeria. In: Sonderforschungsbereich 268 'Kultur- und Sprachgeschichte im Naturraum Westafrikanische Savanne' (Ed.): *Proceedings. International Symposium 27.-29.5.1999, Frankfurt/Main*. Frankfurt a. M., SFB 268 an der Universität Frankfurt. In Press.

Kudhongania A. and A. J. Cordone 1974. Past trends, present stocks and possible future state of the fisheries of the Tanzanian part of the Lake Victoria *African Journal of Tropical Hydrobiology and Fisheries* 3 1974): 167-182

Kuperan, K. and N. M. R. Abdullah. 1994. Small-scale coastal fisheries and co-management. *Marine Policy* 18 (4): 306-313.

Kutengule, M., 2000. *Farm and Non-Farm Sources of Income: Rural Livelihood Diversification in Malawi*. Unpublished Ph.D. Thesis. Norwich, University of East Anglia.

Kyangwa, M. and Geheb, K. 2000. Regulation in Uganda's Lake Victoria fishery: historical and contemporary conditions. In Geheb, K. and Crean, K. (Eds.) 2000. The Co-management Survey: Co-managerial perspectives for Lake Victoria's fisheries *LVFRP Technical Document* No. 11. LVFRP/TECH/00/11. Jinja, The Socio-economic Data Working Group of the Lake Victoria Fisheries Research Project: 150 - 161.

Kydd, J., 1984. Malawi in the 1970s: Development policies and economic change. Paper to Conference on Malawi: An Alternative Pattern of Development. Centre of African Studies, University of Edinburgh, 24-25 May, 1984.

Kydd, J. and A. Hewitt, 1986. The effectiveness of structural adjustment lending: initial evidence from Malawi. *World Development* 14: 347-365.

Lawry, S. W. 1994. Structural adjustment and natural resources in the Sub-Saharan Africa: the role of tenure reform. *Society and Natural Resources*, 7 (1994): 383-387.

Leach, M. Mearns, R. and Scoones, I. 1999. Environmental entitlements: dynamics and institutions in community based natural resources management. *World Development* 27 (2): 225 - 247.

Lele, U., 1990. Structural adjustment, agricultural development and the poor: some lessons from the Malawi experience. *World Development* 18: 1207-1219.

Levieil, D. P. and B. Orlove, 1990. Local control of aquatic resources: community and ecology in Lake Titicaca, Peru. *American Anthropologist* 92 (2): 362-382.

Lewis, D. S. C. and D. Tweddle, 1990. The yield of usipa (*Engraulicypris sardella*) from the Nankumba Peninsula, Lake Malawi (1985-1986). In: Pitcher, T. J. and C. E. Hollingworth (Eds). *Collected Reports on Fisheries Research in Malawi. Occasional Papers, Volume 1*. London, Overseas Development Administration: 57-66.

Lewis, W. A., 1954. Economic development with unlimited supplies of labour. *The Manchester School* 22: 139-191

Lindley, R., 2000. Lake Tanganyika Biodiversity Project Fishing Practices Special Study. UNDP/GEF/RAF/92/G32.

Lowe, R. H., 1952. *Report on the Tilapia and Other Fish and Fisheries of Lake Nyasa. Part II 1945-47*. London, Crown Agents for the Colonies.

Lowe-McConnell, R. H., 1969. Speciation in tropical freshwater bodies. In *Biol. J. Linn. Soc.* 1: 51-75.

Lowe-McConnell, R. H., 1987. *Ecological studies in tropical fish communities*. Cambridge, Cambridge Univ. Press, 382 pp

Lowe-McConnell, R H, 1993. Fish faunas of the African Great Lakes: Origins, diversity and vulnerability. *Conservation Biology* 7: 634-643.

Lowore, J. and J. Lowore. 1999. Community management of natural resources in Malawi. State of the Environment Study. GOM/DANIDA Lake Chilwa Wetland and Catchment Management Project (Unpublished), Zomba, Malawi 10: 45-46.

LTBP (Lake Tanganyika Biodiversity Project), 2000a. Lake Tanganyika: The transboundary diagnostic analysis. UNDP/GEF/RAF/92/G32.

LTBP (Lake Tanganyika Biodiversity Project), 2000b. The first Strategic Action Programme for the sustainable development of Lake Tanganyika. UNDP/GEF/RAF/92/G32. (July 2000).

LTBP (Lake Tanganyika Biodiversity Project), 2000c. The Convention on the sustainable development of Lake Tanganyika. UNDP/GEF/ RAF/92/ G32. (Working Draft No. 4).

Ludwig, D., R. Hilborn, and C. Walters, 1993. Uncertainty, resource exploitation, and conservation. *Science* 260 (April 2, 1993): p.17 and p.36.

Maembe, T. W., 1996. Report on Fisheries Management and Institutions in the countries bordering Lake Tanganyika. FAO/FINNIDA Research for the Management of the Fisheries of Lake Tanganyika. GCP/RAF/271/FIN-TD/57 (En):43p.

Magnet, C., J. E. Reynolds, and H. Bru, 2000. Lake Tanganyika Regional Fisheries Programme: A proposal for the implementation of the Lake Tanganyika Regional Framework Fisheries Management Plan. FAO/ FISHCODE, GCP/INT/648/NOR, Field Report F-14 (En): 128p. Rome, FAO.

Mamdani, M. 1996. *Citizen and subject: contemporary Afric and the legacy of late colonialism*. New Jersey, Princeton University Press.

Mannini, P., 1998. Geographical distribution patterns of pelagic fish and macrozooplankton in Lake Tanganyika. FAO/FINNIDA Research for the Management of the Fisheries on Lake Tanganyika. GCP/RAF/271/FIN-TD/83 (En): 125 pp.

Mannini, P., 1999. Lake Tanganyika Fisheries Monitoring Programme. FAO/ FINNIDA Research for the Management of the Fisheries of Lake Tanganyika. GCP/RAF/271/FIN-TD/90 (En): in print.

Maphosa, F.1996. *The role of kinship in indigenous business in Zimbabwe*. Unpublished D.Phil. thesis, Department of Sociology, University of Zimbabwe.

Mapila, S. A., K. Nyiranda and E. M. Makawa, 1998. *A Guide to the Fisheries Conservation and Management Act, 1997*. Government of Malawi: Fisheries Department Bulletin No 36.

Marks, S. A. 1991 Some reflections on participation and co-management from Zambia's Central Luangwa Valley. In: West, P. and Brechin, S. (Eds). *Resident people and national parks: social dilemmas and strategies in international conservation*. Tucson, University of Arizona Press: 347-358.

Marshall, B. E. Junor F. J. R. and Langerman, J. D. 1982. *Fisheries and Fish Production on the Zimbabwean Side Of Lake Kariba*. Kariba Studies, Paper Number 10. Department of National Parks, Harare, Zimbabwe

Martens, K., G. W. Coulter and B. Goddeeris (Eds.), 1994. Speciation in Ancient Lakes. *Arch. Hydrobiol. Beih. Ergebn. Limnol* 44: 508 pp.

Masundire, H. M. 1997. Spatial and temporal variations in the composition and density of Crustacean plankton in five basins of Lake Kariba Zambia - Zimbabwe. *Journal of Plankton Research* 19 (1): 43 - 62.

McCay, B., 1993. Management regimes. Paper presented to the Property Rights and the Performance of Natural Resource Systems Workshop held on the 2-4 September, 1993 at the Beijer International Institute of Ecological Economics, Stockholm.

McCay, B. J. and Acheson, J. M. (Eds) 1987. *The question of the commons: the culture and ecology of communal resources*. Tucson, University of Arizona Press.

McCracken, J., 1987. Fishing and the colonial economy: the case of Malawi. *Journal of African History* 28: 413-429.

McEvedy, C. 1995. *The Penguin Atlas of African History*. London, Penguin Books.

Matthes, H., 1967. Preliminary investigations into the biology of the Lake Tanganyika Clupeidae. *Fisheries Research Bulletin of Zambia* 4: 39-45.

Medard, M. 1996. Report of the socio-economic study of the fisheries of Mwanza and Speke Gulfs following the ban of the use of non-selective gear. Mwanza, Tanzania, Tanzania Fisheries Research Institute.

Medard, M. 2000a. Community-based organisation on Lake Victoria: a lesson from Tweyambe fishing enterprise in Muleba District, Kagera Region, Tanzania. In K. Geheb, and K. Crean, (Eds.) 2000. The Co-management Survey: Co-managerial perspectives for Lake Victoria's fisheries *LVFRP Technical Document* No. 11. LVFRP/TECH/00/11. Jinja, The Socio-economic Data Working Group of the Lake Victoria Fisheries Research Project: 94 - 106.

Medard, M. 2000b. Community Based Organisation: Lessons from Tweyambe Fishing Enterprise in Lake Victoria, Tanzania. Paper presented at the Workshop on Gender, Globalization and Fisheries held at the Memorial University of Newfoundland, Canada, May 6 - 13, 2000.

Medard, M. and K. Geheb, 2000. Fisheries management in the social domain: perspectives from Tanzania's Lake Victoria fishery. In Geheb, K. and Crean, K. (Eds.) 2000. The Co-management Survey: Co-managerial perspectives for Lake Victoria's fisheries *LVFRP Technical Document* No. 11. LVFRP/TECH/00/11. Jinja, The Socio-economic Data Working Group of the Lake Victoria Fisheries Research Project: 116 – 134.

Medard, M., Mlahagwa, E., Kabati, M., Komba, D., Msunga, D. and Ngusa, D. 2000. Report of the PRA carried out at Mwasonge Beach, Tanzania, September 2-10, 2000. In Geheb, K. (Ed.) The Co-management Survey: PRA reports from five beaches on Lake Victoria. *LVFRP Technical Document* No. 9. LVFRP/TECH/00/9. The Socio-economic Data Working Group of the Lake Victoria Fisheries Research Project, Jinja: 147 - 182.

Meadows, K. and K. Zwick, 2000. Socio-economic special study: final report. UNDP/GEF/RAF/92/G32. 55p.

Meeren, A. G. L. van der 1980. *A socio-anthropological analysis of the fisheries of lake Chad*. Report No FI:DP/NIR/74/001. Rome, Food and Agricultural Organization.

Menz, A., 1995. *The Fishery Potential and Productivity of the Pelagic Zone of Lake Malawi/Niassa*. Greenwich, UK, Natural Resources Institute.

Menz, A. and A. B. Thompson, 1995. *Management Report of the UK/SADC Pelagic Fish Resource Assessment Project, Lake Malawi/Niassa*. London, Overseas Development Administration.

Mitullah, W. L V. 1998. Lake Victoria's Nile perch fish industry: the politics of joint action. *IDS Working Paper* No. 519. Institute of Development Studies, University of Nairobi.

Mkumbo, O. C. and I.G. Cowx, 1999. Catch trends from Lake Victoria – Tanzanian waters. In Cowx, I. G. and Tweddle, D. (Eds.) Report of the fourth FIDAWOG workshop held at Kisumu 16 to 20 August 1999. *LVFRP Technical document* No. 7. LVFRP/TECH/99/07. Jinja, Fisheries Data Working Group of the Lake Victoria Fisheries Research Project: 99 – 122.

Moinuddin, H. 1991. Workshop on Conflict within the Kapenta Industry. Zambia-Zimbabwe SADC Fisheries Project. Workshop Report held at Lake View in Kariba, September 16th and 17th, 1991. ZZSFP, Kariba, Zimbabwe

Mölsä, H., J. E. Reynolds, E. J. Coenen, and O. V. Lindqvist, 1999. Fisheries research towards resource management on Lake Tanganyika. *Hydrobiologia* 407: 1 –24.

Mölsä, H., Sarvala, J. Badende, S. Chitamwebwa, D. Kanyaru, R. Mulimbwa, M. and L. Mwape, 2001. Ecosystem monitoring in the development of sustainable fisheries in Lake Tanganyika. *Aquatic Ecosystem Health & Management* (in press).

Moreau, J.1997. Main hydrological characteristics of Lake Kariba. In: Moreau, J. (Ed.) *Advances in the ecology of Lake Kariba*. Harare, Zimbabwe, University of Zimbabwe Publications.

Msosa, W., 1999. Fishery culture and origins of the Tonga people of Lake Malawi. In: Kawanabe, H., G. W. Coulter and A. C. Roosevelt (Eds), *Ancient Lakes: Their Cultural and Biological Diversity*. Ghent, Belgium, Kenobi Productions: 271-280.

Mtika, B. 1996. Fisheries co-management activities, achievements and constraints. In Fisheries Department (compiled by F. Njaya). Progress Review of the Participatory Fisheries Management Programme for Lake Malombe and Upper Shire River. Proceedings of a workshop held at Boadzulu Lakeshore Resort, Mangochi, 27-29 August, 1996). Lilongwe, Malawi, Fisheries Department: 72-81.

Murindagamo, F., 1992. Wildlife management in Zimbabwe: the CAMPFIRE programme. *Unasylva* 168 (43): 20-26.

Murphree M. W. 1991. *Communities as institutions for resource management*. Harare, Centre for Applied Social Sciences, University of Zimbabwe.

Murphree, M. W., G. Cheater, B.J Dorsy, and B.D. Monthobi, 1975. *Education, race and employment in Rhodesia*. Salisbury, Rhodesia, Artca Publication.

Murray, S. N., R. F. Ambrose, J. A. Bohnsack, L. W. Botsford, M. H. Carr, G. E. Davis, P. K. Dayton, D. Gotshall, D. R. Gunderson, M. A. Hixon, J. Lubcenco, M. Mangel, A. MacCall, D. A. McArdle, J. C. Ogden, J. Roughgarden, R. M. Starr, M. J. Tegner, N. M. Yoklavich, 1999. No-take reserve networks: sustaining fishery populations and marine ecosystems. *Fisheries* 24 (11): 11-25.

NARMAP, 1999. Plan of Operation: National Aquatic Resource Management Programme. Malawi Department of Fisheries/GTZ. Lilongwe, Malawi, April 1999.

National Economic Council. 1998. Malawi Government Economic Report 1998 (Budget Document No. 4). Zomba, Government Printer.

Neiland, A. E. and I. Verinumbe. 1990. Fisheries development and resource-usage conflict: a case-study of deforestation associated with the Lake Chad Fishery in Nigeria. *Environmental Conservation* 18 (2): 111-117.

Neiland, A. E. and M.-T. Sarch. 1993. *The development of a survey methodology for the investigation of traditional management of artisanal fisheries, North East Nigeria.* CEMARE Report no R24

Njaya, F. J., Donda, S. J. and M. M. Hara. 1999. Fisheries co-management: a case study of Lake Chiuta, Malawi. Paper presented at the ICLARM/IFM International workshop on Fisheries Co-management from 23-28 August, 1999, Penang, Malaysia.

Njaya, F. and S. Chimatiro, 1999. *Technology Strategies for the Fisheries and Fish Farming Sustainable Livelihoods Systems in Malawi.* Lilongwe, Malawi Industrial Research and Technology Development Centre (MIRTDC)/UNDP Sustainable Livelihoods Programme. Discussion Paper, November 1999.

N'jie, M. and Mikkola, H. 2001. A Fisheries Co-management Case Study from The Gambia. *Naga* 24 (3&4) (in press).

Noble, B. F. 2000. Institutional criteria for co-management. *Marine Policy* 24 (2000): 69-77.

Normann, A. K., J. R. Nielsen and S. Sverdrup-Jensen 1998. *Fisheries Co-management in Africa: Proceedings from a regional workshop on fisheries co-management research.* Hirtshals, Denmark, Institute for Fisheries Management and Coastal Community Development: Fisheries Co-management Research Project Research Report Number 12.

Oakerson, R. J. 1992. Analysing the commons: a framework. In Bromley, D. W. (Ed.). *Making the commons work: theory, practice and policy.* San Francisco, Institute for Contemporary Studies Press: 41 - 59.

Okemwa, E. 1995. The management and directions for future research on Lake Victoria Multispecies Fisheries. In Okemwa, E., Wakwabi, E. and A. Getabu, (Eds.). *Recent trends of research on Lake Victoria fisheries.* Proceedings of the Second EEC Regional Seminar on Recent Trends of Research on Lake Victoria Fisheries, 25-27th September, 1991, Kisumu, Kenya. Kenya Marine and Fisheries Research Institute and the European Economic Community. Nairobi, ICIPE Science Press: 183-194.

Olivier de Sardan, J. P. 1999. A moral economy of corruption in Africa? *The Journal of Modern African Studies* 37 (1): 25-52.

Olivry, J.-C., A. Chouret, G. Vuillaume, J. Lemoalle and J.-P. Bricquet. 1996. *Hydrologie du lac Tchad*. Paris, ORSTOM Editions.

Onyango, P. O. 2000. Ownership and co-management: towards an integrated co-management of Lake Victoria. In Geheb, K. and Crean, K. (Eds.) 2000. The Co-management Survey: Co-managerial perspectives for Lake Victoria's fisheries *LVFRP Technical Document* No. 11. LVFRP/TECH/ 00/11. Jinja, The Socio-economic Data Working Group of the Lake Victoria Fisheries Research Project: 107 -115.

Ostrom, E. 1990 *Governing the Commons: The evolution of institutions for collective action*. New York, Cambridge University Press.

Ostrom, E. 1995a. Constituting social capital and collective action. *Journal of Theoretical Politics* 6 (4): 527-562.

Ostrom E. 1995b. Designing complexity to govern complexity. In: Hanna S., and Munasinghe M. (Eds). *Property rights and the environment: social and ecological issues*. Washington: The Beijer International Institute of Ecological Economics and the World Bank, Washington: 33 - 47.

Ostrom, E. 1996. Crossing the great divide: coproduction, synergy and development. *World Development* 24 (6): 1073-1087.

Ostrom, E. 1997. A behavioral approach to the rational-choice theory of collective action. *American Political Science Review* 92 (1): 1-22.

Oye, K. A. and Maxwell, J. H. 1995. Self-interest and environmental management. In Keohane, R. O. and Ostrom, E. (Eds.) *Local commons and global interdependence: heterogeneity and cooperation in two domains*. London, Sage Publications: 191-122.

Paffen, P., E. Coenen, S. Bambara, M. Wa Bazolana, E. Lyimo and C. Lukwesa, 1997. Synthesis of the 1995 simultaneous frame survey of Lake Tanganyika Fisheries. FAO/FINNIDA Research for the Management of the Fisheries on Lake Tanganyika. GCP/RAF/271/FIN-TD/60 (En): 35p.

Patterson, G., 2000. Sedimentation Special Study: Summary of findings. UNDP/GEF/RAF/92/G32. 21p.

Pearce, M. J., 1985a. A description and stock assessment of the pelagic fishery in the south-east arm of the Zambian waters of Lake Tanganyika. Report to the Department of Fisheries, Zambia. Mimeo, 74p.

Pearce, M. J., 1985b. The deepwater demersal fish in the south of Lake Tanganyika. Report to the Department of Fisheries, Zambia. Mimeo. 163p.

Pearce, M. J., 1985c. Some effects of *Lates* species on pelagic and demersal fish in Zambian waters of Lake Tanganyika. In: D. Lewis (Ed.), *CIFA Occasional Papers, 15*: 69 –87. Rome, FAO.

Pearce, M., 1995. The removal and return of fishermen in the Sinazongwe Islands. Unpublished manuscript. Sinazongwe, Zambia/Zimbabwe SADC Fisheries Project.

Petit, P. and A. Kiyuku, 1995. Changes in the pelagic fisheries of northern Lake Tanganyika during the 1980s. In: Pitcher, T. J. and Hart, P. J. B (Eds) *The Impact of Species Changes in African Lakes*. London, Chapman and Hall: 443-456.

Petr, T. and J. M. Kapetsky, 1983. Pelagic fish and fisheries of tropical and subtropical lakes and reservoirs. *NAGA (ICLARM Quarterly)* 6 (3): 9-11.

Pinkerton, E. 1989. Introduction: attaining better fisheries management through co-management - prospects, problems and propositions. In Pinkerton, E. (Ed) *Co-operative management of local fisheries* Vancouver: University of British Columbia Press: 3 – 36.

Pitcher, T. J. and P. J. B. Hart (Eds.), 1995. *The Impact of Species Changes in African Lakes*. London, Chapman and Hall.

Platteau, J-P. 1992. Small-scale fisheries and the evolutionist theory of development. In Tvedten, I. and Hersoug, B. (Eds.). *Fishing for development: small-scale fisheries in Africa*. Uppsala, Scandinavian Institute for African Studies: 91-114.

Plisnier, P.-D., 1997. Climate, limnology and fisheries changes of Lake Tanganyika. FAO/FINNIDA Research for the Management of the Fisheries of Lake Tanganyika. GCP/RAF/271/FIN-TD/72 (En): 39p..

Pomeroy, R. S., 1994. Introduction. In R. S. Pomeroy (Ed.), *Community management and common property of coastal fisheries in Asia and the Pacific: concepts, methods and experiences*. Manila, International Centre for Living Aquatic Resources Management: 1-11.

Pomeroy, R. S., and F. Berkes, 1997. Two to tango: the role of government in fisheries co-management. *Marine Policy* 21 (5): 465-480.

Pomeroy, R. S. and M.J. Williams 1994. Fisheries co-management and small-scale fisheries: a policy brief. Manila, The Philippines: International Center for Living Aquatic Resource Management.

Pomeroy, S. R., B. M. Katon and I. Harkes. Conditions Affecting the Success of Fisheries Co-management: lessons from Asia. Forthcoming in *Marine Policy*

Pullin, R. 1991. Foreword. In: Msiska, O. V. and B. A. Costa-Pierce (Eds). *History, Status and Future of Common Carp (*Cyprinus carpio *L.) as an Exotic Species in Malawi.* ICLARM, Philippines: 1.

Quan, J., 1993. *Socio-economic Research Strategies for the Fisheries Sector in Malawi.* Greenwich, Natural Resources Institute.

Rahman, M.A 1993. Glimpses of the 'Other Africa'. *People's Self – Development: Perspectives on Participatory Action Research. A journey through experience.* Dhaka, University Press Limited.

Redclift, M. 1984. *Development and the environmental crisis: red or green alternatives?* London, Methuen.

Redclift, M. 1987. *Sustainable development: exploring the contradictions.* London, Routledge.

Reynolds, J. E., 1998. Regional framework planning for Lake Tanganyika fisheries management. FAO/FINNIDA Research for the Management of the Fisheries of Lake Tanganyika. GCP/RAF/271/FIN-TD/89 (En): 69p.

Reynolds, J. E., 1999a. Building management partnerships: local referenda on fisheries futures for Lake Tanganyika. FAO/FINNIDA Research for the Management of the Fisheries of Lake Tanganyika. GCP/RAF/271/FIN-TD/91 (En): 92p.

Reynolds, J. E., 1999b. Lake Tanganyika Framework Fisheries Management Plan. FAO/FISHCODE, GCP/INT/648/NOR, Field Report F-2 (En): 43 p. Rome, FAO.

Reynolds, J. E., (Ed.) 1997a. LTR lake-wide socio-economic survey, 1997: Zambia. (Co-Authors: P. Paffen, E. Bosma, V. Langenburg, R. Chitembure, J. Chimanga, W. Chomba, C. Lukwesa, and M. Mwenda.) FAO/FINNIDA Research for the Management of the Fisheries of Lake Tanganyika. GCP/ RAF/271/FIN-TD/67 (En).

Reynolds, J. E. (Ed.), 1997b. LTR lake-wide socio-economic survey, 1997: Tanzania. (Co-Authors: E. Bosma, P. Paffen, P. Verburg, D. B. R. Chitamwebwa, K. I. Katonah, F. Sob, A. N. M. Kalangali, L. Nonde, S. Muhoza, and E. Kadula.) FAO/FINNIDA Research for the Management of the Fisheries of Lake Tanganyika. GCP/RAF/271/FIN-TD/68 (En).

Reynolds, J. E. (Ed.), 1997c. LTR lake-wide socio-economic survey, 1997: Democratic Republic of Congo. (Co-Authors: E. Bosma, P. Paffen, N. Mulimbwa, G. Kitungano, C. Nyiringabi, A. Kwibe, C. Bulambo, E. Mukirania, and I. Mbilizi.) FAO/FINNIDA Research for the Management of the Fisheries of Lake Tanganyika. GCP/RAF/271/FIN-TD/69 (En).

Reynolds, J. E. (Ed.), 1997d. LTR lake-wide socio-economic survey, 1997: Burundi. (Co-Authors: P. Paffen, E. Bosma, J. M. Tumba, C. Butoyi, E. Gahungu, E. Nikomeze, and B. Ndimunzigo.) FAO/FINNIDA Research for the Management of the Fisheries of Lake Tanganyika. GCP/RAF/271/ FIN-TD/70 (En).

Reynolds, J. E. and G. Hanek, 1997. Tanganyika fisheries and local stakeholders. An Overview of the LTR Lake-wide Socio-economic Survey, 1997. FAO/FINNIDA Research for the Management of the Fisheries of Lake Tanganyika. GCP/RAF/271/FIN-TD/71 (En): 72p.

Reynolds, J. E. and D.F. Greboval, 1988. Socio-economic effects of the evolution of Nile Perch fisheries in Lake Victoria: a review. *CIFA Technical Paper* No. 17. Rome, Committee for the Inland Fisheries of Africa, Food and Agricultural Organization.

Reynolds J. E., D. Gréboval and P. Mannini 1992. *Thirty years on: observation on the development of the Nile Perch Fishery in Lake Victoria* Bujumbura: UNDP/FAO Regional Project for Inland Fisheries Planning (IFIP).

Reynolds, J.E. and H. Mölsä, 2000. Lake Tanganyika Regional Fisheries Programme (TREFIP) – Environmental Impact Assessment Study. FAO/ FISHCODE, GCP/INT/648/NOR, Field Report F-15 (En): 94 p. Rome, FAO.

Reynolds, J. E. and P. Paffen, 1997. LTR lake-wide socio-economic survey, 1997: Notes on methods and procedures. FAO/FINNIDA Research for the Management of the Fisheries of Lake Tanganyika. GCP/RAF/ 271/ FIN-TD/66 (En): 79p.

Rhodesian Herald, 26 August 1975. Fish industry needs single authority, Salisbury, Rhodesia, *The Rhodesian Herald.*

Riedmiller, S.1994. Lake Victoria fisheries: the Kenyan reality and environmental implications. *Environmental Biology of Fishes* 39 (1994): 329-338.

Roberts, C. M., 1998. Sources, sinks, and the design of marine reserve networks. *Fisheries* 23 (7): 16-19.

RoU (Republic of Uganda), 1995. *Constitution of the Republic of Uganda.* Published by Authority of the Government of Uganda. Entebbe, Government Printer.

RoU (Republic of Uganda), 1997. *The Local Governments Act, 1997.* Entebbe, Government Printer.

RoU (Republic of Uganda), 1999. *Statistical Abstract. Entebbe, Uganda Bureau of Statistics.*

Salonen, K. Sarvala, J. Järvinen, M. Langenberg, V. Nuottajärvi, M. Vuorio, K and D. Chitamwebwa, 1999. Phytoplankton in Lake Tanganyika – vertical and horizontal distribution of *in vivo* fluorescence. *Hydrobiologia* 407: 89-103.

Sanyanga, R. A., 1996. Variations in abundance of *Synodontis Zambezensis* (Pisces: Mochokidae) Peters 1852 in the inshore fishery of Lake Kariba. *Fisheries Research* 26 (1996): 171-186

Sarch, M.-T. 1999. *Fishing and farming at Lake Chad: A livelihood analysis.* Ph.D. Thesis, School of Development Studies, University of East Anglia, UK.

Sarch M-T., 2000. Institutional Evolution at Lake Chad: Traditional Administration and Flexible Fisheries Management. Proceedings of the Tenth Conference of the International Institute for Fisheries Economics and Trade, Corvallis, Oregon, 10 - 14 July, 2000. http://osu. orst.edu/dept/ IIFET/2000/papers/sarch2.pdf.

Sarch, M-T. and E. H. Allison, 2000. Fluctuating fisheries in Africa's inland waters: well-adapted livelihoods, maladapted management. Proceedings of the 10th International Conference of the Institute of Fisheries Economics and Trade. Corvallis, Oregon, 9-14 July 2000. http://osu.orst.edu/dept/IIFET/2000/papers/sarch.pdf.

Sarch M-T. and C. M. Birkett. 2000. Fishing and farming at Lake Chad: responses to lake level fluctuations. *The Geographical Journal* 163 (2): 1 - 17

Sarvala, J., K. Salonen, M. Järvinen, E. Aro, T. Huttula, P. Kotilainen, H. Kurki, V. Langenberg, P. Mannini, A. Peltonen, P-D. Plisnier, I. Vuorinen, H. Mölsä and O. V. Lindqvist, 1999. Trophic structure of Lake Tanganyika: carbon flows in the pelagic food web. *Hydrobiologia* 407: 149 – 173.

Scudder, T., 1960. Fishermen of the Zambezi: an appraisal of fishing practice and potential of the valley Tonga. *Rhodes-Livingstone Journal* 27 (1960): 41-49.

Scudder, T. and T. Conelly, 1985. Management systems for riverine fisheries. *FAO Fisheries Technical Paper* No. 263. Rome, Food and Agricultural Organization.

Scudder, T., and J. Habarad, 1991. Local responses to involuntary relocation and development in the Zambian portion of the middle-Zambezi Valley. In J. A. Mollet (Ed.). *Migrants in Development*. London, Macmillan: 178-205.

SEDAWOG, 1999a. Marketing survey. *LVFRP Technical Document* No. 2. LVFRP/TECH/99/02. Jinja, Socio-economic Data Working Group of the Lake Victoria Fisheries Research Project.

SEDAWOG, 1999b. The survey of Lake Victoria's fishers. *LVFRP Technical Document* No. 5. LVFRP/TECH/99/05. Jinja, Socio-economic Data Working Group of the Lake Victoria Fisheries Research Project.

SEDAWOG, 2000. Fisheries co-management options at Kiumba Beach: a participatory pilot study. *LVFRP Technical Document* No. 8. LVFRP/TECH/00/08. Jinja, Socio-economic Data Working Group of the Lake Victoria Fisheries Research Project.

Sen, S. and J.R. Nielsen, 1996. Fisheries co-management: a comparative analysis. *Marine Policy* 20 (5): 405-418.

Siachoono, S. M., 1995. Contingent valuation as an additional tool for evaluating wildlife utilization management in Zambia: Mumwa Game Management Area. *AMBIO* 24 (4): 246- 249.

Sikes, S.K. 1972. *Lake Chad*. London, Eyre Methuen.

Sinazongwe District Council, 1995. Sinazongwe District Council Secretary's Report to the Special Local Authority Meeting to be held on 11/9/1995 at 10:00 hrs. Sinazongwe, Sinazongwe District Council.

Smale, M., 1995. Maize is life: Malawi's delayed green revolution. *World Development* 23: 819-831.

Songore, N., Moyo, A. and Chitembure, R. M. 2000. 1999 *Joint Fisheries Statistical Reports Lake Kariba*. Kariba, Zimbabwe, Zambia-Zimbabwe SADC Fisheries Project.

State Enterprises and Indigenisation Department. 1998. *Indigenisation of the economy of Zimbabwe,* Policy Paper Series No. 4. Harare, Zimbabwe, State Enterprises and Indigenisation Department.

State Enterprises and Indigenisation Department. 1999. *Indigenisation of the economy of Zimbabwe: Mechanisms to co-ordinate strategies for implementation of Indigenisation Programme of Action at National, Provincial and Sectoral* levels. Harare, Zimbabwe, State Enterprises and Indigenisation Department.

Swallow, B. M., Meinzen-Dick, R. S., Jackson, L. A. Williams, T. O. and Anderson White, T. 1997. Multiple functions of common property regimes. Panel presented at International Association for the Study of Common Property; 6th Annual Conference, June 7th, 1997. *EPTD Workshop Summary Paper* No. 5. Environment and Production Technology Division, International Food Policy Research Institute, Washington D. C., U. S. A

Tarbit, J., 1972. *Lake Malawi Trawling Survey – Interim Report 1969-71*. Fisheries Bulletin, No. 3. Zomba, Malawi: Extension Aids Section, Ministry of Agriculture and Natural Resources.

Tellegen, N., 1997. *Rural Enterprises in Malawi: Necessity or Opportunity?* Research Series No. 12/1997. Avebury, UK: Ashgate.

Temple, O. 1919. *Notes on the tribes, provinces, emirates and states of the northern provinces of Nigeria* Cape Town, Argus.

Temple, P. H. 1965. Physical factors influencing land use in coastal Sukumaland. *East African Geographical Review* 3 (April 1965): 17 – 26.

Thomasson, T. and M. C. Banda, 1996. *Depth Distribution of Fish Species in Southern Lake Malawi – Implications for Fisheries Management*. Fisheries Bulletin No. 34. Lilongwe, Malawi: Department of Fisheries.

Thompson, A. B. and E. H. Allison, 1997. Potential yield estimates of unexploited pelagic fish stocks in Lake Malawi. *Fisheries Management and Ecology* 4: 31-48.

Turner, G. F., 1995. Management, conservation and species changes of exploited fish stocks in Lake Malawi. In: Pitcher, T. J. and Hart, P. J. B (Eds.) *The Impact of Species Changes in African Lakes*. London, Chapman and Hall: 365-396.

Turner, G. F., 1996. Maximisation of yields from African lakes. In: Cowx, I (Ed.). *Stock Assessment in Inland Fisheries*. Oxford, Fishing News Books, Blackwell Science: 465-481.

Turner, M. D. 1999. Conflict, environmental change, and social institutions in dryland Africa: limitations of the community resource management approach. *Society and Natural Resources*, 12 (1999): 643-657.

Tweddle, D. and J. Magasa, 1989. Assessment of multispecies cichlid fisheries of the south-east arm of Lake Malawi, Africa. *J. Cons. int. Explor. Mer.* 45 (1989): 209 – 222.

Uganda Government 1964. *The Fish and Crocodiles Act* Chapter 228, Revised Edition 1964. Government Printer, Entebbe, Uganda.

UNDP/FAO, 1994. Project of the Government of Malawi: component document. (Unpublished): 3-4

UNDP/FAO, 1995. Chambo Fisheries Research, Malawi: project findings and recommendations. Food and Agriculture Organization, Rome: 25-35.

UNDP/GEF, 2001. Project document (PDF-B): RAF01G41/A/1G/31 Developing detailed regional and national project proposals and financing mechanisms to implement the Lake Tanganyika Strategic Action Program.

Vanden Bossche, J-P. and Bernacsek, G. M. 1990. Source book for inland fishery resources of Africa. *CIFA Technical Paper* No. 18. 3 vol.s. Committee for the Inland Fisheries of Africa. Rome, Food and Agricultural Organization.

van Jaarsveld, A. S. and S. L. Chown, 2001. Climate change and its impacts in South Africa. *TREE* 16 (1): 13-14.

Walter, G., 1988. Lake Kariba (Zambia): socio-economic baseline study. Unpublished consultancy report to the Ministry of Agriculture and Water Development, Lusaka.

Walters, C. J. and C.S. Holling, 1990. Large-scale management experiments and learning by doing. *Ecology* 71 (6): 2037-2068.

Walters, C. J. and R. Hilborn, 1978. Ecological optimization and adaptive management. *Annual Review of Ecology and Systematics* 9 (1978): 157-188.

Waters, J. R. 1991. Restricted access vs open access methods of management: toward more effective regulations of fishing effort. *Marine Fisheries Review*. 53 (3): 1 – 10.

Watson, C. E. P, 1987. *Final Report: Malawi Fisheries Development Strategy Study*. Lilongwe, Malawi: Unpublished report to Department of Fisheries, Government of Malawi, by GOPA Consultants Ltd.

WCED. 1987. *Our common future.* Oxford, OUP for the World Commission on Environment and Development.

Welcomme, R. L. 1979. *Fisheries ecology of floodplain rivers*. London, Longman.

West, K. (Ed.), 2001. Lake Tanganyika: Results and experiences of the UNDP/GEF Conservation initiative. UNDP/GEF/RAF/92/G32. 138p.

Wilson, D. C. 1999. The Global in the Local: The Environmental State and the Management of the Nile Perch Fishery on Lake Victoria. Paper Presented at the RC 24 Miniconference on The Environmental State Under Pressure: The Issues and the Research Agenda, August 6th and 7th Northwestern University, Evanston IL

Wilson, D.C. and M. Medard, n.d. Country fieldwork report of the socioeconomic research on Lake Victoria, Tanzania by the Program on the Lakes of East Africa 1994 – 1995. Consisting of excerpts from D.C Wilson 1996. The critical human ecology of the Lake Victoria fishing industry. Michigan State University: Dissertation, and with an introduction by Modesta Medard, Tanzania Fisheries Research Institute.

Wilson, D. C., Medard, M. Harris, C. K. and Wiley, D. S. 1996. *Potentials for comanagement of the Nile perch fishery – Lake Victoria, Tanzania.* Paper presented at the 'Voices from the Commons' conference, International Association for the study of common property, Berkeley, CA, June 5-8, 1996.

Wilson, D. C., Medard, M. Harris, C. K. and Wiley, D. S. 1999. The implications for participatory fisheries management of intensified commercialization on Lake Victoria. *Rural Sociology* 64 (4): 554-572.

Wilson, J., 1993. Lake Malombe and Upper Shire River community fisheries management programme. *SADC Natural Resources Newsletter*, 4: 10-11.

Wilson, J. A. 1982. The Economical Management of Multispecies Fisheries, *Land Economics*, 58 (4): 417-434.

Wilson, J. A., J. M. Acheson, M. Metcalfe and P. Kleban, 1994. Chaos, complexity and community management of fisheries. *Marine Policy* 18 (4): 300-301.

Witte, F., Goldschmidt, A., Goudswaard, P. C., Ligtvoet, W., Van Oijen, M. J. P. and Wanink, J. H. 1992. Species extinction and concomitant ecological changes in Lake Victoria. *Netherlands Journal of Zoology* 42 (2-3), Ch. 26: 214-232.

World Bank, 1999. Country profiles (Burundi, Democratic Republic of Congo, Tanzania, Zambia). Web page: http://www.worldbank.org/html/extdr/ offrep/ afr (as at 10 March 1999).

World Bank, 2000. *Lake Malawi/Nyasa Environmental Management: Project Concept Document.* Africa Regional Office, Southern and Eastern Africa Rural Development Operations.

World Commission on Environment and Development. 1987. *Our common future*. Oxford and New York, Oxford University Press.

Worthington, E. B., 1996. Early Research on East African Lakes: An Historical Sketch. In: Johnson, T. C. and E. O. Odada (Eds.), *The Limnology, Climatology and Paleoclimatology of the East African Lakes*. Newark, New Jersey, Gordon and Breach.

Zimbabwe Independent. 13 June 1997. Supreme Court rules on kapenta fishing. Harare, Zimbabwe, *Zimbabwe Independent*.

ZZSFP (Zambia-Zimbabwe SADC Fisheries Project), 1989. *Project Proposal*. Unpublished document.

ZZSFP (Zambia/Zimbabwe SADC Fisheries Project), 1995. Minutes of the 3rd Finance Sub-Committee of the Lake Kariba Sinazongwe Interim Zonal Management Committee held on 14th November, 1995. Sinazongwe, Department of Fisheries.

ZZSFP (Zambia/Zimbabwe SADC Fisheries Project), 1997a. Minutes of the 5th Inter Zonal Committee Meeting held at Sinazongwe Fisheries Training Centre, Sinazongwe, 25 and 26th June 1997. Sinazongwe, Zambia/Zimbabwe SADC Fisheries Project.

ZZSFP (Zambia-Zimbabwe SADC Fisheries Project), 1997b, Working group on bio-economic Assessment of Kapenta (Limnothrissa miodon) in Lake Kariba (Zambia and Zimbabwe), Project Report Number 50.

Index

www.ingramcontent.com/pod-product-compliance
Lightning Source LLC
Chambersburg PA
CBHW060336220326
41598CB00023B/2728

* 9 7 8 9 9 7 0 0 2 2 9 3 9 *